Group Theory for High Energy Physicists

Group Theory for High Energy Physicists

Mohammad Saleem
Muhammad Rafique

CRC Press is an imprint of the
Taylor & Francis Group, an **informa** business
A TAYLOR & FRANCIS BOOK

CRC Press
Taylor & Francis Group
6000 Broken Sound Parkway NW, Suite 300
Boca Raton, FL 33487-2742

First issued in paperback 2019

ISBN-13: 978-1-4665-1063-0 (hbk)
ISBN-13: 978-0-367-38089-2 (pbk)

Visit the Taylor & Francis Web site at
http://www.taylorandfrancis.com

and the CRC Press Web site at
http://www.crcpress.com

Contents

Preface.....ix
About the Author.....xi

1 Elements of Group Theory.....1
1.1 Definition of a Group.....1
1.2 Some Characteristics of Group Elements.....4
1.3 Permutation Groups.....6
1.4 Multiplication Table.....10
1.5 Subgroups.....10
1.6 Power of an Element of a Group.....13
1.7 Cyclic Groups.....14
1.8 Cosets.....16
1.9 Conjugate Elements and Conjugate Classes.....17
1.10 Conjugate Subgroups.....17
1.11 Normal Subgroups.....18
1.12 Center of a Group.....18
1.13 Factor Group.....19
1.14 Mapping.....20
1.15 Homomorphism.....22
1.16 Kernel.....24
1.17 Isomorphism.....25
1.18 Direct Product of Groups.....27
1.19 Direct Product of Subgroups.....29

2 Group Representations.....31
2.1 Linear Vector Spaces.....31
2.2 Linearly Independent Vectors.....33
2.3 Basis Vectors.....33
2.4 Operators.....34
2.5 Unitary and Hilbert Vector Spaces.....35
2.6 Matrix Representative of a Linear Operator.....36
2.7 Change of Basis and Matrix Representative of a Linear Operator.....40
2.8 Group Representations.....44
2.9 Equivalent and Unitary Representations.....47
2.10 Reducible and Irreducible Representations.....48
2.11 Complex Conjugate and Adjoint Representations.....49
2.12 Construction of Representations by Addition.....49
2.13 Analysis of Representations.....51
2.14 Irreducible Invariant Subspace.....52

2.15 Matrix Representations and Invariant Subspaces......52
2.16 Product Representations......57

3 Continuous Groups......61
3.1 Definition of a Continuous Group......61
3.2 Groups of Linear Transformations......62
3.3 Order of a Group of Transformations......69
3.4 Lie Groups......72
3.5 Generators of Lie Groups......75
3.6 Real Orthogonal Group in Two Dimensions: O(2)......84
3.7 Generators of SU(2)......91
3.8 Generators of SU(3)......95
3.9 Generators and Parameterization of a Group......98
3.10 Matrix Representatives of Generators......99
3.11 Structure Constants......101
3.12 Rank of a Lie Group......103
3.13 Lie Algebras......104
3.14 Commutation Relations between the Generators of a Semisimple Lie Group......105
3.15 Properties of the Roots......108
3.16 Structure Constants $N_{\alpha\beta}$......111
3.17 Classification of Simple Groups......112
3.18 Roots of SU(2)......114
3.19 Roots of SU(3)......115
3.20 Numerical Values of the Structure Constants of SU(3)......122
3.21 Weights of a Representation......122
3.22 Computation of the Highest Weight of Any Irreducible Representation of SU(3)......127
3.23 Dimension of any Irreducible Representation of SU(N)......131
3.24 Computation of the Weights of Any Irreducible Representation of SU(3)......133
3.25 Weights of Irreducible Representation $D^8(1,1)$ of SU(3)......135
3.26 Weight Diagrams......138
3.27 Decomposition of a Product of Two Irreducible Representations......139
3.27.1 First Method......139
3.27.2 Second Method......141

4 Symmetry, Lie Groups, and Physics......147
4.1 Symmetry......147
4.1.1 Rotational Symmetry......147
4.1.2 Higher and Lower Symmetries......151
4.1.3 Reflection/Inversion Symmetry......151
4.1.4 Concept of Parity......153

4.1.5 Multiple Symmetries ... 155
4.1.6 Combination of Symmetry Operations ... 155
4.1.7 Translational Symmetry in Space ... 156
4.1.8 Time-Reversal Symmetry ... 157
4.1.9 Charge Conjugation ... 159
4.1.10 Symmetry Groups and Physics ... 162
4.2 Casimir Operators ... 165
4.3 Symmetry Group and Unitary Symmetry ... 166
4.4 Symmetry and Physics ... 166
4.5 Group Theory and Elementary Particles ... 170
Reference ... 190

Appendix A ... 191

Appendix B ... 195

Appendix C ... 199

Appendix D ... 203

Index ... 207

Preface

Group theory has played an exciting, fascinating, and significant role in the development of various disciplines of physics. However, it is amazing that during the last three decades no book has been written that starts *ab initio* and then builds on and considers applications for group theory from the point of view of high energy physicists. This book is an attempt to achieve this objective.

Chapter 1 introduces the concept of a *group* and then presents the characteristics that are imperative for developing group theory as it applies to high energy physics. Chapter 2 describes group representations, as physicists have always found it more convenient to deal with matrix representations of a group than the abstract group itself. Group representations are important because various irreducible invariant subspaces of a group can be used to accommodate elementary particles with specific characteristics. Chapter 3 discusses continuous groups, which have given rise to spectacular progress in our understanding of particles and their interactions. The subject, however, has not been developed in a mathematically rigorous manner. The aim of this book is to introduce the concept of continuous groups, especially Lie groups, and their characteristics in a way that is easily comprehensible to physicists. The root structure of some important groups is analyzed, and the weights of various representations of these groups are obtained. All three chapters are interspersed with examples and problems. These chapters can therefore form the content of a group theory course for undergraduate students interested in specializing in high energy physics.

In the first three chapters, group theory is developed to a level that it can be applied to solve high energy physics problems. Chapter 4 shows how symmetry principles associated with group theoretical techniques can be used to interpret some experimental results and make predictions. The appendices at the end of the book prove some important relations and theorems given in the text.

It is hoped that the book will be useful to undergraduate as well as graduate students in physics and mathematics and researchers in high energy physics.

Dr. Muhammad Rafique, coauthor of the book, breathed his last while the book was still in preparation. May his soul rest in peace.

About the Author

Dr. Mohammad Saleem obtained his BA (Hons), MA (math), MSc (physics), and PhD from University of the Punjab and his BSc (Special Hons) from the London University. He has been professor and chair of the Department of Physics, founder and director of the Centre for High Energy Physics, professor in the School of Physical Sciences and dean of the Faculty of Science at University of the Punjab. At present, he is professor emeritus at the University of the Punjab. He is also professor at the Institute for Basic Research in Palm Harbor, Florida. He has written more than 150 research papers on high energy physics, most of which have been published in standard foreign journals or were accepted for presentation at international conferences. His books on special relativity and high energy physics, written in collaboration with his colleagues, have been published in the United Kingdom and the United States. He is also an editor of the *Hadronic Journal*.

Dr. Muhammad Rafique graduated from University of the Punjab in 1962. He obtained his MSc and PhD from University of North Wales. He spent a year at the International Centre for Theoretical Physics on a postdoctoral fellowship and then worked for five years as associate professor of applied mathematics at Alfateh University in Tripoli, Libya. He was appointed professor of applied mathematics at University of the Punjab in 1983. He has written more than 50 research papers on high energy physics.

1

Elements of Group Theory

This chapter introduces the concept of a group and presents the characteristics that are imperative for developing group theory as it applies to high energy physics.

1.1 Definition of a Group

What is a group? This can best be answered by reference to a few examples. Consider the set $S = \{1, -1, i, -i\}$. The numbers in curly brackets are called its *elements* or *members*. With the law for the combination of elements, called the *law of composition* or the *binary operation*, as a multiplication of complex numbers, the elements of S possess the following four properties:

1. If an element of S is multiplied, according to the prescribed law of composition, by an element of the same set, the resulting number is again an element of S. This is called the *closure property*. For example,

$$1 \times i = i, \; i \times (-i) = 1, \; (-1) \times (-1) = 1$$

2. The multiplication is *associative*; that is, the product does not depend on the order in which the elements are multiplied. Thus, if a, b, and c are arbitrary elements of S, then $a \times (b \times c) = (a \times b) \times c$. For example:

$$1 \times \{(-1) \times i\} = 1 \times (-i) = -i$$

and

$$\{1 \times (-1)\} \times i = (-1) \times i = -i$$

so that

$$1 \times \{(-1) \times i\} = \{1 \times (-1)\} \times i$$

3. The set S contains an element that, when multiplied by any one of its elements, either from the left or from the right, *reproduces* that element. The element 1 in S possesses this characteristic:

$$1 \times 1 = 1,\ 1 \times (-1) = -1,\ i \times 1 = i,\ 1 \times (-i) = -i$$

   Such an element is said to be the **identity element.**
4. To every element of the set S, there corresponds an element of the same set such that the product of two elements, irrespective of their order, is the identity element. Then any one of these elements is called the *inverse* of the other. For instance, multiplying the elements i and –i of S, we get 1 (i.e., the identity element). By virtue of the previous definition, –i is the inverse of i and i is the inverse of –i.

The set of the cube roots of unity, $\{1,\omega,\omega^2\}$, with $\omega^3 = 1$, also possesses these four characteristics with respect to ordinary multiplication with 1 as the identity and the inverse elements forming the pairs (1, 1), (ω, ω^2), (ω^2, ω).

The set S of two nonsingular matrices I and A given by

$$I = \begin{bmatrix} 1 & 0 \\ 0 & 1 \end{bmatrix} \text{ and } A = \begin{bmatrix} 0 & 1 \\ 1 & 0 \end{bmatrix}$$

exhibits the same four characteristics with the law of composition as matrix multiplication. (A nonsingular matrix is that whose determinant is different from zero.) We notice that:

1. The product of an arbitrary element of the set S with any one of its elements produces an element of this very set. For instance:

$$AA = \begin{bmatrix} 0 & 1 \\ 1 & 0 \end{bmatrix}\begin{bmatrix} 0 & 1 \\ 1 & 0 \end{bmatrix} = \begin{bmatrix} 1 & 0 \\ 0 & 1 \end{bmatrix} = I$$

2. The multiplication of matrices is always associative.
3. I is the identity element because, for instance, IA = A = AI.
4. The inverse of every element exists and is an element of the set. For instance, as AA = I, A is its own inverse.

The fact that different sets of various elements possess the same four characteristics with respect to some given binary operation (law of composition) suggests that they may be given a common name. Such sets are called *groups.*

Thus, a finite or an infinite set G = {a, b, c, …} forms a group with respect to a binary operation, usually called multiplication, if

1. The product of any two elements of G is also an element of G; this includes the product of an element with itself. This is called the *closure property.*
2. The multiplication is associative.
3. The identity element, which occurs when multiplication with any element of the set, either from the right or from the left, reproduces that element, exists, and is an element of G.
4. The inverse of an element, which occurs when multiplication with the element, irrespective of the order, yields the identity element, exists, and is an element of G.

Terms like *product* and *multiplication,* when applied to groups, do not always have their conventional meaning. In the example $(1, \omega, \omega^2)$, the law of combination of elements is multiplication of complex numbers, while in (4) the law of composition is matrix multiplication. The law of composition may be multiplication of numbers, addition of numbers, matrix multiplication, or something else; it is customary to use *product* or *multiplication* for every kind of binary operation. Since in the sequel we shall often have to refer to the elements of the various sets by phrases *belongs to* or *is a member* or *an element of* or *is contained in,* it is convenient to introduce the symbol $\in$, which stands for these phrases. Thus, $a \in G$ means that a belongs to G or is a member of G or is contained in G.

It may be noticed that if we combine the elements of the set S = {1, –1, i, –i} with the addition of complex numbers as the binary operation, the set does not contain the identity element zero and therefore does not form a group.

PROBLEM 1.1

Show that *under addition of complex numbers,* the set $\{1, \omega, \omega^2\}$ does not form a group.

To crystallize our ideas, we recast the definition of a group as follows. A finite or infinite nonempty set G = {a, b, c, …} is said to form a group *under a specified law of composition* if

(1) For every $a, b \in G$, the product $ab \in G$.
(2) For every $a, b, c \in G$, $a(bc) = (ab)c$.
(3) There exists an element $e \in G$ such that $ae = a = ea$ for every $a \in G$. The element e is called the identity element.
(4) For every $a \in G$, there exists an element $a^{-1} \in G$, known as the inverse of a, such that $aa^{-1} = e = a^{-1}a$.

Sometimes postulates (3) and (4) are replaced by weaker postulates that the left identity and the left inverse exist. From these weaker postulates, (3) and (4) can be derived.

A group is said to be *finite* if it has a finite number of elements. Otherwise, it is said to be *infinite*. The number of elements of a finite group is called the *order of the group*. Thus, the finite group $G = \{1, -1, i, -i\}$ is of order 4. The group of all integers negative, zero, and positive, that is, $G = \{\ldots, -2, -1, 0, 1, 2, \ldots\}$ is an infinite group under addition. Such a group is said to be an *additive group*.

The group elements do not necessarily commute with respect to the underlying law of composition. If all the elements of a group G commute with one another, that is, $ab = ba$ for all $a,b \in G$, the group is said to be *commutative* or *Abelian*. Otherwise, it is said to be noncommutative or non-Abelian. In all the examples given above, the groups are Abelian.

PROBLEM 1.2
Give an example of a noncommutative finite group.

PROBLEM 1.3
Is an additive group always Abelian?

1.2 Some Characteristics of Group Elements

We will now show that a group contains only one identity element and there is only one inverse of each element. In other words, in every group, *identity element is unique*, and to each element of the group there corresponds a *unique inverse*. These results can be proved as follows:

1. Suppose that a group contains two identity elements e and e′. Then, by the very definition of an identity element, we must have $ee' = e$ and $ee' = e'$ so that $e = e'$. This shows that a group can contain only one identity element: identity element of a group is unique.
2. We will next show that inverse of any element of a group is also unique. Consider an arbitrary element a of a group G. Suppose that the element a has two inverses, b and c. By definition, both the inverses belong to G and must be such that $ba = ab = e$ and $ac = ca = e$. Multiplying $ba = e$ from the right by c, we get $bac = ec = c$. Similarly, multiplying $ac = e$ from the left by b, we get $bac = be = b$. Comparing the equations $bac = c$ and $bac = b$, we obtain $c = b$. That is, the two inverses of a must be equal. This shows that inverse of an element is unique.
3. It is clear from the definition of the inverse of an element a of a group that if a^{-1} is the inverse of a, then a is the inverse of a^{-1}. Therefore, we may write $(a^{-1})^{-1} = a$.
4. We next prove that the inverse of the product of any two elements of a group is equal to the product of the inverses of those elements taken in the reverse order, that is, $(ab)^{-1} = b^{-1}a^{-1}$ for all $a, b \in G$.

To prove this, we note that since the multiplication is associative, we must have

$$(ab)(b^{-1}a^{-1}) = abb^{-1}a^{-1} = aea^{-1} = aa^{-1} = e$$

This equation shows that $b^{-1}a^{-1}$ is the inverse of ab: $b^{-1}a^{-1} = (ab)^{-1}$. The result can be generalized to any finite number of factors.

PROBLEM 1.4
Show that if a, b, c belong to a group G such that ab = ac, then b = c.

PROBLEM 1.5
Prove that if a is an arbitrary element of a group G, then aG = G, where aG stands for the set of all the elements obtained by multiplying each element of G from the left by a.

PROBLEM 1.6
A rational number can be expressed as a ratio m/n of two integers, m and n, where $n \neq 0$. Show that the set of all positive rational numbers is a group under ordinary multiplication. Can the set of all negative rational numbers form a group under ordinary multiplication?

PROBLEM 1.7
Show that the set of all rational numbers is a group under ordinary addition.

PROBLEM 1.8
Prove that the set of all nonzero rational numbers is a group under ordinary multiplication.

Example 1.1

PROBLEM

Show that the set of all nonsingular 2×2 matrices forms a group under matrix multiplication.

SOLUTION

Consider the set of all nonsingular 2×2 matrices

$$A = \begin{bmatrix} a_{11} & a_{12} \\ a_{21} & a_{22} \end{bmatrix}, \quad B = \begin{bmatrix} b_{11} & b_{12} \\ b_{21} & b_{22} \end{bmatrix}, \ldots$$

Then

(i) The product AB of two arbitrary members A and B of the set is also a square matrix of order 2. Denoting it by C, we have

$$AB = C$$

Taking the determinant of both sides, we get

$$\det(AB) = \det A \cdot \det B = \det C$$

Since A and B are nonsingular, $\det A \neq 0$, $\det B \neq 0$. Therefore, the previous equation shows that $\det C \neq 0$. That is, the 2×2 matrix C is also nonsingular and hence a member of the set.

(ii) The multiplication, being matrix multiplication, is associative.

(iii) The nonsingular 2×2 unit matrix I serves as the identity element.

(iv) Since the matrices are nonsingular, the inverse of each matrix exists and is nonsingular. Moreover, as the inverse of every matrix is of the same order as the matrix itself, its order is 2. Hence, the inverse of every matrix, being a nonsingular and 2×2 matrix, is a member of the set.

Since the set of all nonsingular 2×2 matrices satisfies all the group postulates, it forms a group under matrix multiplication.

1.3 Permutation Groups

Consider the set of all permutations of digits 1, 2, 3, that is, 123, 132, 213, 231, 312, 321.

We may write these as

$$I = \begin{pmatrix} 1 & 2 & 3 \\ 1 & 2 & 3 \end{pmatrix},\ T_1 = \begin{pmatrix} 1 & 2 & 3 \\ 1 & 3 & 2 \end{pmatrix},\ T_2 = \begin{pmatrix} 1 & 2 & 3 \\ 2 & 1 & 3 \end{pmatrix}$$

$$T_3 = \begin{pmatrix} 1 & 2 & 3 \\ 2 & 3 & 1 \end{pmatrix},\ T_4 = \begin{pmatrix} 1 & 2 & 3 \\ 3 & 1 & 2 \end{pmatrix},\ T_5 = \begin{pmatrix} 1 & 2 & 3 \\ 3 & 2 & 1 \end{pmatrix}$$

The permutation T_1 takes 1 into 1, 2 into 3, and 3 into 2. We can therefore write T_1 also as

$$\begin{pmatrix} 1 & 3 & 2 \\ 1 & 2 & 3 \end{pmatrix}$$

or

$$\begin{pmatrix} 2 & 1 & 3 \\ 3 & 1 & 2 \end{pmatrix}$$

since these permutations also take 1 into 1, 2 into 3, and 3 into 2. That is, the permutation is unchanged by shifting the elements in the first row if the corresponding elements in the second row are also shifted in the same manner. We shall show that the set of permutations I, T_1, T_2, T_3, T_4, T_5 forms a group *under successive operation of permutations from right to left*. For instance, the *product* $T_1 T_2$ of two permutations T_1 and T_2 is defined as the permutation obtained by carrying out first the permutation T_2 and then the permutation T_1.

(i) Now

$$T_1T_2 = \begin{pmatrix} 1 & 2 & 3 \\ 1 & 3 & 2 \end{pmatrix}\begin{pmatrix} 1 & 2 & 3 \\ 2 & 1 & 3 \end{pmatrix}$$

This may be written as

$$T_1T_2 = \begin{pmatrix} 2 & 1 & 3 \\ 3 & 1 & 2 \end{pmatrix}\begin{pmatrix} 1 & 2 & 3 \\ 2 & 1 & 3 \end{pmatrix}$$

This shows that T_2 changes 1 2 3 to 2 1 3 and T_1 changes 2 1 3 to 3 1 2. The net effect is that 1 2 3 is changed to 3 1 2. Hence,

$$T_1T_2 = \begin{pmatrix} 1 & 2 & 3 \\ 3 & 1 & 2 \end{pmatrix} = T_4$$

which is a member of the set. In fact, we can show that the product of any two members of the set is also a member of the set.

(ii) The multiplication is associative. We will show it for one case only:

$$T_1(T_2T_3) = (T_1T_2)T_3$$

We have

$$T_1(T_2T_3) = T_1\begin{pmatrix} 1 & 2 & 3 \\ 2 & 1 & 3 \end{pmatrix}\begin{pmatrix} 1 & 2 & 3 \\ 2 & 3 & 1 \end{pmatrix}$$

$$= T_1\begin{pmatrix} 2 & 3 & 1 \\ 1 & 3 & 2 \end{pmatrix}\begin{pmatrix} 1 & 2 & 3 \\ 2 & 3 & 1 \end{pmatrix}$$

$$= T_1\begin{pmatrix} 1 & 2 & 3 \\ 1 & 3 & 2 \end{pmatrix} = \begin{pmatrix} 1 & 2 & 3 \\ 1 & 3 & 2 \end{pmatrix}\begin{pmatrix} 1 & 2 & 3 \\ 1 & 3 & 2 \end{pmatrix}$$

$$= \begin{pmatrix} 1 & 3 & 2 \\ 1 & 2 & 3 \end{pmatrix}\begin{pmatrix} 1 & 2 & 3 \\ 1 & 3 & 2 \end{pmatrix} = \begin{pmatrix} 1 & 2 & 3 \\ 1 & 2 & 3 \end{pmatrix}$$

and

$$(T_1T_2)T_3 = \begin{pmatrix} 1 & 2 & 3 \\ 1 & 3 & 2 \end{pmatrix}\begin{pmatrix} 1 & 2 & 3 \\ 2 & 1 & 3 \end{pmatrix}T_3$$

$$= \begin{pmatrix} 2 & 1 & 3 \\ 3 & 1 & 2 \end{pmatrix}\begin{pmatrix} 1 & 2 & 3 \\ 2 & 1 & 3 \end{pmatrix}T_3$$

$$= \begin{pmatrix} 1 & 2 & 3 \\ 3 & 1 & 2 \end{pmatrix}T_3$$

$$= \begin{pmatrix} 1 & 2 & 3 \\ 3 & 1 & 2 \end{pmatrix}\begin{pmatrix} 1 & 2 & 3 \\ 2 & 3 & 1 \end{pmatrix}$$

$$= \begin{pmatrix} 2 & 3 & 1 \\ 1 & 2 & 3 \end{pmatrix}\begin{pmatrix} 1 & 2 & 3 \\ 2 & 3 & 1 \end{pmatrix} = \begin{pmatrix} 1 & 2 & 3 \\ 1 & 2 & 3 \end{pmatrix}$$

This shows that the multiplication is associative.

(iii) The permutation

$$I = \begin{pmatrix} 1 & 2 & 3 \\ 1 & 2 & 3 \end{pmatrix}$$

is the identity element. This can be easily verified. For instance,

$$IT_1 = \begin{pmatrix} 1 & 2 & 3 \\ 1 & 2 & 3 \end{pmatrix}\begin{pmatrix} 1 & 2 & 3 \\ 1 & 3 & 2 \end{pmatrix}$$

$$= \begin{pmatrix} 1 & 3 & 2 \\ 1 & 3 & 2 \end{pmatrix}\begin{pmatrix} 1 & 2 & 3 \\ 1 & 3 & 2 \end{pmatrix} = \begin{pmatrix} 1 & 2 & 3 \\ 1 & 3 & 2 \end{pmatrix} = T_1$$

and

$$T_1I = \begin{pmatrix} 1 & 2 & 3 \\ 1 & 3 & 2 \end{pmatrix}\begin{pmatrix} 1 & 2 & 3 \\ 1 & 2 & 3 \end{pmatrix} = \begin{pmatrix} 1 & 2 & 3 \\ 1 & 3 & 2 \end{pmatrix} = T_1$$

(iv) Since the correspondence can always be reversed, the inverse of each permutation exists. Let us calculate the inverse of T_4. Let T_i be the inverse of T_4. Then the product of T_4 and T_i must be equal to the identity element *I*:

$$T_i T_4 = \begin{pmatrix} 1 & 2 & 3 \\ a & b & c \end{pmatrix} \begin{pmatrix} 1 & 2 & 3 \\ 3 & 1 & 2 \end{pmatrix} = I = \begin{pmatrix} 1 & 2 & 3 \\ 1 & 2 & 3 \end{pmatrix}$$

or

$$\begin{pmatrix} 3 & 1 & 2 \\ c & a & b \end{pmatrix} \begin{pmatrix} 1 & 2 & 3 \\ 3 & 1 & 2 \end{pmatrix} = \begin{pmatrix} 1 & 2 & 3 \\ 1 & 2 & 3 \end{pmatrix}$$

or

$$\begin{pmatrix} 1 & 2 & 3 \\ c & a & b \end{pmatrix} = \begin{pmatrix} 1 & 2 & 3 \\ 1 & 2 & 3 \end{pmatrix}$$

This yields a = 2 , b = 3, c = 1 so that

$$T_i = \begin{pmatrix} 1 & 2 & 3 \\ a & b & c \end{pmatrix} = \begin{pmatrix} 1 & 2 & 3 \\ 2 & 3 & 1 \end{pmatrix} = T_3$$

Thus, the inverse of T_4 is T_3 which is a member of the set. Similarly, we can show that I, T_1, T_2, T_4, T_5 are, respectively, the inverses of I, T_1, T_2, T_3, T_5. Hence, all the inverses are members of the set of permutations.

The previous analysis shows that the set of permutations I, T_1, T_2, T_3, T_4, T_5 forms a group and that the law of composition is the application of permutations from right to left.

In general, the set of all permutations

$$T = \begin{pmatrix} 1 & 2 & 3 & \cdots & n \\ i_1 & i_2 & i_3 & \cdots & i_n \end{pmatrix}$$

of n objects forms a group. This group is called the *permutation group* or the *symmetric group* and is denoted by S_n. Since the number of permutations of n objects is n!, there are n! transformations and hence n! elements in the group.

1.4 Multiplication Table

Under a given law of composition, the group properties of any set of elements are determined if we know the result of multiplication for each ordered pair of elements. For a finite group, it is convenient to collect all this information by arranging the products of different elements in a tabular form. This is shown as follows for the groups $\{1, \omega, \omega^2\}$ and $\{1, -1, i, -i\}$, where the law of composition is ordinary multiplication. These tables are called multiplication tables. In a multiplication table, the product ab of an element a with an element b appears at the intersection of the row and column headed by a and b, respectively.

	1	ω	ω^2
1	1	ω	ω^2
ω	ω	ω^2	1
ω^2	ω^2	1	ω

	1	–1	i	–i
1	1	–1	i	–i
–1	–1	1	–i	i
i	i	–i	–1	1
–i	–i	i	1	–1

Three of the group axioms can be verified immediately by looking at the table:

1. The fact that the multiplication table contains only the elements listed in the extreme left column and the top-most row shows that the group is closed under multiplication.
2. The existence of the identity element is established by the presence of a row/column identical with the top/side.
3. The fact that the identity element appears only once in every row and column implies the existence of a unique inverse for every element.

The associative law alone requires an elaborate calculation to verify it.

The multiplication tables given here are symmetrical about their leading diagonal. This symmetry implies that ab = ba for all elements of the group, that is, the groups are Abelian.

1.5 Subgroups

Consider a group G = {1, –1, i, –1}, with the law of composition as ordinary multiplication. H = {1, –1} is a subset of this group. This subset forms a group under the same law of composition as for the group G and is called a *subgroup* of G. In general terms, if a subset H of a group G itself forms a group under the same law of composition as for G, then the subset H is said to be a subgroup of G. It may be emphasized that a subset H of G forms a subgroup of G only if H is a group under the *same binary operation* as for G. Thus, although all the elements of the subset H = {1, –1} of the additive group G = {..., –2, –1, 0, 1, 2, ...} belong to G, this subset H is not a subgroup of G, because it does not form a group under ordinary addition.

It may be noticed that the identity element is always a subgroup of G. The identity element and the group G itself are called *improper* or *trivial subgroups*. All other subgroups are called *proper* or *nontrivial subgroups*. Thus, a group always has two improper subgroups: the identity element and the group itself. Normally, the term subgroup will stand for a proper subgroup.

We shall now prove an important theorem that determines whether a subset of a group forms its subgroup.

Theorem 1.1

A subset H of a group G is a subgroup of G if and only if whenever a,b belong to H, the product of a and b^{-1}, viz., ab^{-1} also belongs to H.

PROOF

(i) The condition is necessary.

That is, we have to show that if H is a subgroup of G and $a,b \in H$, then ab^{-1} must belong to H. Since H is a subgroup of G and $b \in H$, the inverse of b (i.e., b^{-1}) not only exists but also belongs to H and consequently the product of $a \in H$ and $b^{-1} \in H$; that is, ab^{-1} should belong to H.

(ii) The condition is sufficient.

Now, we have to show that if any two elements a and b belong to a subset H of a group G such that the product ab^{-1} also belongs to H, then the subset H forms a subgroup of G. To prove this, we first notice that since all the elements of the subset also belong to the group G, their inverses must exist. Consider two elements $a,b \in H$. Then, it is given that $ab^{-1} \in H$. By choosing b = a, we have $aa^{-1} \in H$ or $e \in H$. That is, H contains the identity element. Now, since the identity $e \in H$ and for any element $a \in H$, it follows from the given condition that $ea^{-1} = a^{-1} \in H$. This means that the inverse of any element a of H also belongs to H. Consequently, if a and b are any two elements of H, then, by virtue of the previous result, a^{-1} and b^{-1} should belong to H. But, $a,b^{-1} \in H$ implies that $a(b^{-1})^{-1} = a\,b \in H$; that is, the set H is closed under multiplication. Finally, the associativity follows because it holds in G. Hence, H is a subgroup of G.

The order of a subgroup is related to the order of its group by Lagrange's theorem.

Lagrange's Theorem

The order of a subgroup H of a finite group G is a divisor of the order of G.

PROOF

Let G = {a, b, c, ...} be a finite group of order n. Suppose that H = {h_1, h_2, ..., h_m} is a subgroup of G and is of order m. If m = n, the theorem is proved since then m (= n) is a divisor of n. If m ≠ n, the group is not exhausted. Then there must exist an element a of G that does not belong to H. We form the product aH = {ah_1, ah_2, ..., ah_m}. We make the following observations about the elements of aH:

(i) As a ∈ G and H is a subgroup of G, all the elements of aH should belong to G. This is so because as every element of H belongs to G and as the element a also belongs to G, the product of a with any element of H should also belong to G.
(ii) All the elements of aH are distinct. This follows because if $ah_j = ah_k$, for j ≠ k, then $a^{-1}ah_j = a^{-1}ah_k$. This yields $h_j = h_k$. This contradicts the initial assumption that j ≠ k. Hence the elements of aH are all distinct so that aH contains exactly m elements.
(iii) The subgroup H and the set aH do not have a common element. If possible, for a ∈ G but not ∉ H, let $ah_j = h_\ell$ for some j and ℓ. For ℓ = j, we have $ah_j = h_j$. This is possible only if a is the identity element. Since the identity element must belong to the subgroup H, this is against the initial assumption that a does not belong to H. If j ≠ ℓ, then multiplying equation $ah_j = h_\ell$ by h_j^{-1} from the right, we get $ah_jh_j^{-1} = h_\ell h_j^{-1}$. This gives $a = h_\ell h_j^{-1} = h_m$, say, because h_ℓ and h_j are elements of H. This shows that a is an element of H. This contradicts the initial assumption that a is not an element of H. Hence it is not possible to have $ah_j = h_\ell$. This discussion shows that the set aH contains m distinct elements of G, none of which is common with H. We have thus separated 2m elements of G. If it exhausts the group G, then n = 2m (i.e., n/m = 2) and the theorem is proved. If the group is not yet exhausted, there must exist another element b of G such that b is contained in neither H nor aH. Let us form the set bH = {bh_1, bh_2, ..., bh_m}. As before, all the elements of bH are distinct, and the set bH does not have any element in common with H. We can also show that bH has no element in common with aH. If possible, let $bh_k = ah_\ell$ for some k and ℓ. Then $bh_kh_k^{-1} = ah_\ell h_k^{-1}$ or $b = ah_s$ for some s. Therefore, b ∈ aH. This is against the initial assumption that b does not belong to aH. Hence, the elements of bH are distinct from the elements of H and aH. If this exhausts the group G, we must have n/m = 3, and the theorem is proved. If the group is still not exhausted, we can proceed further along the same lines. Since the

group is of finite order, it must be exhausted after a finite number of steps. Thus, we have $n = km$, where k is a positive integer, or $n/m = k$. This proves Lagrange's theorem. The integer k is called the *index* of H in G.

For instance, the order of the group {1, –1, i, –i} is 4. The order of its subgroup {1, –1} is 2, which divides 4. It follows that a group of prime order has no nontrivial subgroup.

PROBLEM 1.9
Find all six subgroups of the symmetric group S_3.

1.6 Power of an Element of a Group

Let a be an arbitrary element of a group G. Then, by the very definition of a group, a a, a a a, ..., being the products of a group element with itself, also belong to G. These products are written as $a^2, a^3, \ldots$. In general, if in the product $aa \cdots a$, then the element a occurs k times as a factor and we write the product as a^k and call it the k-th power of a. Thus, a^k is the element that results by writing k times the element a and multiplying all the a's. Since k can be only a *positive integer*, k = 1, 2, 3, ..., the element a can have only positive integral powers. Similarly, if a^{-1}, the inverse of the element a of G, occurs ℓ times in the product $a^{-1}a^{-1} \cdots a^{-1}$, we write it as $(a^{-1})^\ell$, where ℓ is a positive integer.

We will now prove that $(a^m)^{-1} = (a^{-1})^m$. We have

$$\begin{aligned}(a^m)^{-1} &= (aa \cdots a)^{-1} \text{ m factors}\\ &= a^{-1} \cdots a^{-1}a^{-1} \text{m factors}\\ &= (a^{-1})^m\end{aligned}$$

This gives the desired proof. If we agree to denote each side of the previous equation by a^{-m}, we can write

$$(a^m)^{-1} = (a^{-1})^m = a^{-m}$$

This analysis shows that an element of a group can have only integral powers, positive or negative.

PROBLEM 1.10
Show that for all choices of the integers m and n, the following laws of exponents hold:

$$a^m a^n = a^{m+n}$$

$$(a^m)^n = a^{mn}$$

1.7 Cyclic Groups

Consider a group G = {a, b, c, …}, which may or may not be finite. Suppose that all the elements of G can be created by taking the powers of an appropriately chosen element, say a, of G. Then G is called a *cyclic group,* and a is called a *generator* of the group. For instance, consider the element i of a finite group G = {1, –1, i, –i} with the law of composition as ordinary multiplication. We notice that as $i^1 = i$, $i^2 = -1$, $i^3 = -i$, $i^4 = 1$, the whole group is generated by taking only the positive powers of the element i. If we take higher powers of i, elements of the group start repeating themselves. Thus, it is a cyclic group. The element i is a generator of the group. We may generate the same group by taking only negative powers of i. We may also generate this group by taking positive powers of –i. The generator of a cyclic group is therefore not necessarily unique. Next consider an infinite group of even integers, viz.,

$$\{\ldots, -6, -4, -2, 0, 2, 4, 6, \ldots\}$$

in which the law of composition is ordinary addition. The group can be generated by taking integral (positive and negative) powers of its element 2, which is thus a generator of the group. The element –2 is another generator of this group. Notice that for an infinite cyclic group the powers of its generator are all distinct. We observe that if a is a generator of a cyclic group, then each element of the group has the form a^p, where p is an integer. If a is an element of a group and p is the smallest positive integer for which a^p is the identity element (i.e., $a^p = e$, where e is the identity element), then p is called the *order, period* or *cycle* of the element a. If p is the order of a generator of a *cyclic group,* it is also the *order of the cyclic group.* In the example {1,–1, i,–i}, the order p of the generator i is 4.

PROBLEM 1.11

Show that a cyclic group of order p generated by an element a can also be generated by a^k if k and p are relatively prime.

PROBLEM 1.12

Show that if the order of a cyclic group is prime, then any element (different from the identity) of the group can generate the group.

Consider a *finite* group. All the powers of its arbitrary element a cannot be distinct. Otherwise a, aa, aaa, …, being members of the group, will make its order *infinite.* On the other hand, all the powers of an arbitrary element a of an infinite group G may or may not be distinct. For instance, the order of the identity element of any group is one. If all the powers of a are distinct, then the element a is said to be of *infinite order.*

PROBLEM 1.13
Show that for an element a of order p of a finite group G, all the elements of the set $H = \{a, a^2, \ldots, a^p = e\}$ are distinct and H forms a subgroup of G.

This is a very interesting result because it shows that if a is an arbitrary element of order p of a finite group G, then $\{a, a^2, \ldots, a^p = e\}$ will be a subgroup of G. Since this subgroup is Abelian, it follows that even a non-Abelian group may have proper Abelian subgroups.

Let us illustrate it with an example. The set of matrices

$$A = \begin{bmatrix} \omega & 0 \\ 0 & \omega^2 \end{bmatrix}, \ B = \begin{bmatrix} \omega^2 & 0 \\ 0 & \omega \end{bmatrix}, \ C = \begin{bmatrix} 0 & 1 \\ 1 & 0 \end{bmatrix}$$

$$D = \begin{bmatrix} 0 & \omega^2 \\ \omega & 0 \end{bmatrix}, \ E = \begin{bmatrix} 0 & \omega \\ \omega^2 & 0 \end{bmatrix}, \ I = \begin{bmatrix} 1 & 0 \\ 0 & 1 \end{bmatrix}$$

where ω is a cube root of unity ($\omega^3 = 1$) and forms a non-Abelian group $G = \{A, B, C, D, E, I\}$ of order 6 under matrix multiplication. Consider the element B of this matrix group. Then

$$B^2 = \begin{bmatrix} \omega^2 & 0 \\ 0 & \omega \end{bmatrix}\begin{bmatrix} \omega^2 & 0 \\ 0 & \omega \end{bmatrix} = \begin{bmatrix} \omega & 0 \\ 0 & \omega^2 \end{bmatrix} = A$$

and

$$B^3 = \begin{bmatrix} \omega^6 & 0 \\ 0 & \omega^3 \end{bmatrix} = \begin{bmatrix} 1 & 0 \\ 0 & 1 \end{bmatrix} = I$$

It can be easily verified that the elements B, $B^2 = A$, $B^3 = I$ form a subgroup of order 3. This subgroup consists of the element B and its powers. It is therefore a cyclic subgroup. Moreover, it is an Abelian subgroup because $AB = B^2B = B^3 = I = BA$. The period of B is 3, and it divides the order of the group.

Next, take the element C. Then

$$C^2 = \begin{bmatrix} 0 & 1 \\ 1 & 0 \end{bmatrix}\begin{bmatrix} 0 & 1 \\ 1 & 0 \end{bmatrix} = \begin{bmatrix} 1 & 0 \\ 0 & 1 \end{bmatrix} = I$$

The elements C, $C^2 = I$ form another Abelian cyclic subgroup of G. The order of this subgroup is 2, and it also divides the order 6 of the group.

PROBLEM 1.14

Show that any subgroup of a cyclic group is itself a cyclic group.

1.8 Cosets

Consider a group G. Let H be a subgroup of a group G. Let a be an arbitrary element of G. Then the set of elements ah_i, where $h_i \in H$, is defined as the left coset of H with respect to a and is denoted by aH. The right coset of H with respect to a is defined as the set of elements $h_i a$, where $h_i \in H$ and is denoted by Ha. In general, the left and right cosets aH and Ha are not identical. However, if these are identical (i.e., aH and Ha are equal), it expresses only that the set of elements in Ha is the same as the set of elements in aH. This equivalence does not mean that ah_i is necessarily equal to $h_i a$; that is, the elements of H need not commute with every element of G.

We illustrate it with an example.

Consider the non-Abelian group G of six matrices {A, B, C, D, E, I} with

$$A = \begin{bmatrix} \omega & 0 \\ 0 & \omega^2 \end{bmatrix}, \quad B = \begin{bmatrix} \omega^2 & 0 \\ 0 & \omega \end{bmatrix}, \quad C = \begin{bmatrix} 0 & 1 \\ 1 & 0 \end{bmatrix}$$

$$D = \begin{bmatrix} 0 & \omega^2 \\ \omega & 0 \end{bmatrix}, \quad E = \begin{bmatrix} 0 & \omega \\ \omega^2 & 0 \end{bmatrix}, \quad I = \begin{bmatrix} 1 & 0 \\ 0 & 1 \end{bmatrix}$$

where ω is a cube root of unity ($\omega^3 = 1$), and the law of composition is matrix multiplication. The subset H = {A, B, I} is an Abelian subgroup of G. The left and right cosets of H with respect to C are

$$CH = \{CA, CB, CI\} = \{D, E, C\}$$

$$HC = \{AC, BC, IC\} = \{E, D, C\}$$

We notice that the *collection of elements* of CH is the same as that of HC. It is this fact that is expressed by writing CH = HC. It does not necessarily mean that C commutes with all the elements of H, for instance, $CA \neq AC$.

REMARK

In literature, a different definition of coset is also given: the choice of $a \in G$ is subject to the condition that it does not belong to H.

PROBLEM 1.15
Show that the left cosets aH and bH of the a subgroup H of a group G, where a and b are two arbitrary elements of G, contain either exactly the same elements of G or have no common element. Verify that for G = {1, –1, i, –i} and H = {1, –1}, with the law of composition as ordinary multiplication, the left cosets iH and –iH have exactly the same elements of G.

1.9 Conjugate Elements and Conjugate Classes

Let a and c be two elements of a group G. Then c is said to be *conjugate* to the element a if there exists an element of b of G such that $c = b^{-1}ab$. The elements a and c are said to be *equivalent* elements of G. Further, as $a = e^{-1}\ ae$, any element of a group is conjugate to itself.

PROBLEM 1.16
Show that if a is conjugate to b and b is conjugate to c, then a is conjugate to c.

The set of all elements of a group G that are conjugate to a given element a of G is said to form the *conjugate class* of a. We denote this class by C_a. Since the identity element e of any group always commutes with all the members of the group, we have

$$u^{-1}eu = u^{-1}u = e \quad \text{for all } u \in G$$

This shows that the identity element of any group is a class by itself: C_e = {e}. Similarly, every element of an Abelian group is a class by itself.

We have seen that the identity element of a group forms a class by itself. Therefore, no other class can contain the identity element. Hence, except for the class consisting of the identity, no class can be a subgroup.

1.10 Conjugate Subgroups

Let G = {a, b, c, …} be a group and $H = \{h_1, h_2, h_3, \ldots\}$ a subgroup of G. Let a be an arbitrary element of G. Consider the set

$$K = a^{-1}Ha = a^{-1}\ \{h_1, h_2, h_3, \ldots\}a = \{a^{-1}h_1a, a^{-1}h_2a, a^{-1}h_3a, \ldots\}$$

Since a, a^{-1}, and h_i all belong to G, the elements $a^{-1}h_ia$ of K, being the products of these elements, also belong to G. Thus, K is a subset of G. We will prove that K is a subgroup of G.

Let $\{k_1, k_2, k_3, \ldots\}$ be the elements of K. To prove that K is a subgroup of G, we have to show that if k_i and k_j belong to K, then the element $k_i k_j^{-1}$ also belongs to K. By virtue of the relation $K = a^{-1}Ha$, we have

$$K = \{k_1, k_2, k_3, \ldots\} = \{a^{-1}h_1a, a^{-1}h_2a, a^{-1}h_3a, \ldots\}$$

The elements k_i and k_j of the subset K are therefore related to the elements of the subgroup H as $k_i = a^{-1}h_ia$ and $k_j = a^{-1}h_ja$. Then

$$\begin{aligned} k_ik_j^{-1} &= (a^{-1}h_ia)\,(a^{-1}h_ja)^{-1} \\ &= a^{-1}h_iaa^{-1}h_j^{-1}(a^{-1})^{-1} \\ &= a^{-1}h_ih_j^{-1}a = a^{-1}\,h_la = k_l \in K \end{aligned}$$

Hence, $K = a^{-1}Ha$ is a subgroup of G. This subgroup is said to be the *conjugate subgroup* of H in G with respect to a. The subgroups $a^{-1}Ha$, $b^{-1}Hb$, $c^{-1}Hc$, …, where a, b, c, … all belong to G form the class of subgroups conjugate to H.

1.11 Normal Subgroups

Let H be a subgroup of a group G. If it so happens that for all $a \in G$, the subgroup $a^{-1}Ha$ is the subgroup H itself ($H = a^{-1}Ha$), then H is said to be a *self-conjugate* or *normal* or *invariant subgroup* of G. Multiplying the last equation from the left by a, we get $aH = Ha$. That is, in the case of a normal subgroup, the left and right cosets of H with respect to a are equal. If G is Abelian, all the subgroups of G are normal.

It can be verified that $\{1, -1\}$ is a normal subgroup of $\{1, -1, i, -i\}$.

Obviously, the group G itself and its identity element e are always normal subgroups of G. These are called *trivial normal subgroups.* Other normal subgroups are said to be *nontrivial* or *proper.* Thus, every group has two trivial normal subgroups.

If a group does not contain any proper normal subgroup, it is said to be a *simple* group. All cyclic groups of prime order are simple. The group $1, \omega, \omega^2$ under multiplication of complex numbers is a simple group.

A group that may or may not possess proper normal subgroups but does not possess any Abelian proper normal subgroup is called a *semisimple* group. Every simple group is semisimple.

1.12 Center of a Group

The set Z of all the elements of a group G that commute with every element of G is called the *center of G.*

We will show that Z, the center of G, is a subgroup of G. Let $z \in Z$ and g be an arbitrary element of G. Then, by definition,

$$gz = zg \tag{1.1}$$

As z is an element of the group G, its inverse, z^{-1}, exists. Therefore, multiplying from the right as well as from the left by z^{-1}, we get

$$z^{-1}gzz^{-1} = z^{-1}zgz^{-1}$$

or

$$z^{-1}g = gz^{-1} \tag{1.2}$$

Since g is arbitrary, the previous equation shows that z^{-1} commutes with every element of G and is, therefore, a member of the set Z. Thus, the inverse of every element of Z also belongs to Z.

Now if $z_i, z_j \in Z$ and g is an arbitrary element of G, then

$$z_i z_j^{-1} g = z_i g z_j^{-1} = g z_i z_j^{-1}$$

where, in obtaining this result, Equations 1.1 and 1.2 have been used. This equation shows that $z_i z_j^{-1} \in Z$. In other words, if z_i and z_j are members of the set Z, then $z_i z_j^{-1}$ is also a member of the set. Therefore, Z is a *subgroup of G.*

We notice that as $gZ = Zg$ for all $g \in G$, the subgroup Z is a *normal subgroup of G.*

Any subgroup of Z is also necessarily a subgroup of G. Such a subgroup is called a *central normal subgroup.*

1.13 Factor Group

Let N be a normal subgroup of a group G. Then by definition its right and left cosets are identical; that is, $aN = Na$ for all $a \in G$. Let aN, bN, cN, ..., be the set of left cosets of N. We will show that the set of left cosets forms a group under coset multiplication. The coset multiplication is a process in which a set is obtained by multiplying each element of the first coset with every element of the other, and the repeated elements are counted only once. In general, the product of two cosets will depend on their order.

PROBLEM 1.17

Show that if N is a group, then under coset multiplication, $NN = N$.

The multiplication of cosets is associative because the elements of each left coset belong to the group G for which the multiplication is associative.

The product of two left cosets aN and bN is given by

$$(aN)(bN) = aNbN = abNN = abN = cN$$

where c = ab belongs to G. Therefore, the product of two left cosets is again a left coset.

Since (aN) N = aNN = aN and N (aN) = aNN = aN for *any* left coset aN, the element N is the identity of the set of left cosets with respect to coset multiplication.

If $a \in G$, then a^{-1} also belongs to G. Therefore, for any element $a \in G$, aN and $a^{-1}N$ are left cosets. Then, the product of the cosets aN and $a^{-1}N$ yields

$$(aN)(a^{-1}N) = aa^{-1}NN = eN = N$$

This shows that $a^{-1}N$ is the inverse of aN.

Hence, the set of all left cosets of N forms a group under coset multiplication. This group is called the *factor group* or the *quotient group* of G by N and is denoted by G/N.

Since the number of cosets of N is equal to the index of the subgroup N, it follows that the order of the factor group G/N is equal to the index of N in G.

1.14 Mapping

Consider a set, S, consisting of a mango, a banana, an orange, and an apple and another set, S′, consisting of red, green, and blue baskets, as shown in Figure 1.1. Let φ be a correspondence that associates the mango with the red basket, the banana with the green basket, and both the orange and apple with the blue basket. That is, φ is a *prescription* or *correspondence* that associates each element of S with a unique element of S′ such that no element of S′ is left out. Then φ is called a mapping of the set S onto the set S′.

Next, consider the same two sets of elements, but this time, as shown in Figure 1.2, there is a different correspondence between the elements of the two sets. Each element of S is associated with a unique element of S′,

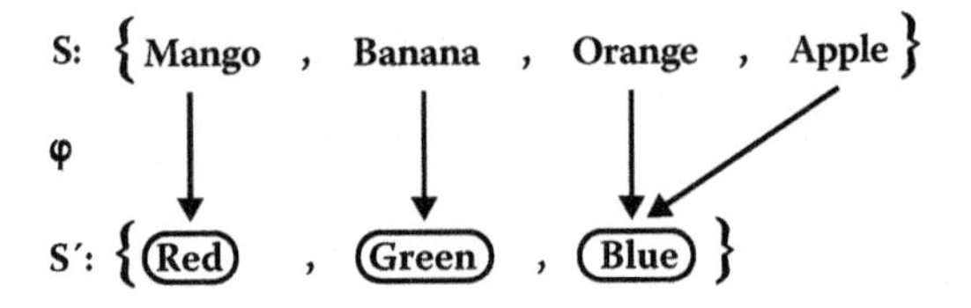

FIGURE 1.1
The mapping of the set S onto the set S′.

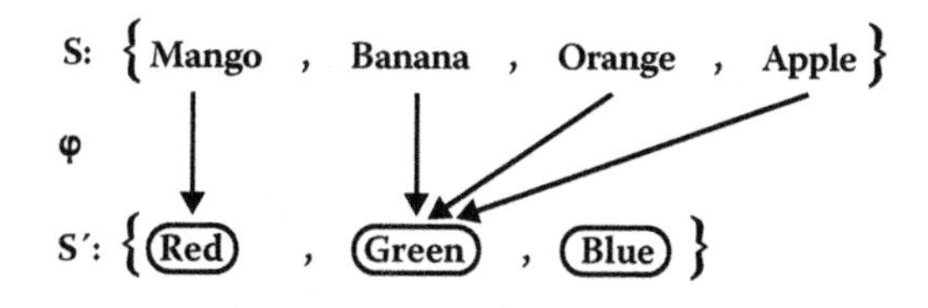

FIGURE 1.2
The mapping of the set S into the set S′.

but in two of three only two elements (red and green baskets) of S′ (i.e., a subset of S′) are associated with the elements of the set S. One basket (blue) is left out. Such a correspondence φ that associates each element of S with a unique element of S′ but where one or more elements of S′ are left out is said to be a mapping of S into S′. Let us generalize. Consider two sets S = {a, b, c, ...} and S′ = {a′, b′, c′, ...}. Suppose that φ is a correspondence that associates each element of S with a *unique* element of S′. Then φ is called a mapping of S to S′. If an element a of S is associated by a mapping φ with an element a′ of S′, then we write it as $a \to a'$ or, more frequently, as $a \to \varphi(a) \equiv a'$. Notice that φ(a) is that element of S′ that is associated with the element a of S. The element a′ is called the *image* of a under the mapping φ. Let φ(S) denote the set of the elements of S′ that correspond to all the elements of S under φ. If φ(S) = S′ (i.e., it is the entire set S′), then φ is said to be a mapping of S onto S′. If φ(S) is a subset of S′, then φ is said to be a mapping of S into S′. Evidently, in the last case, some of the elements of S′ are not associated with any element of S.

If at least one element of S′ corresponds to many elements of S, then φ is said to be *many-to-one mapping*. Thus, in both the previous examples, the mapping is many-to-one. In the first example, it is onto mapping, whereas in the second example it is into mapping. If any element of S′ does not correspond to more than one element of S, then φ is said to be a *one-to-one mapping*. For instance, in the following two examples, the mapping is one-to-one, but it is onto in the first example and into in the second example.

First example:

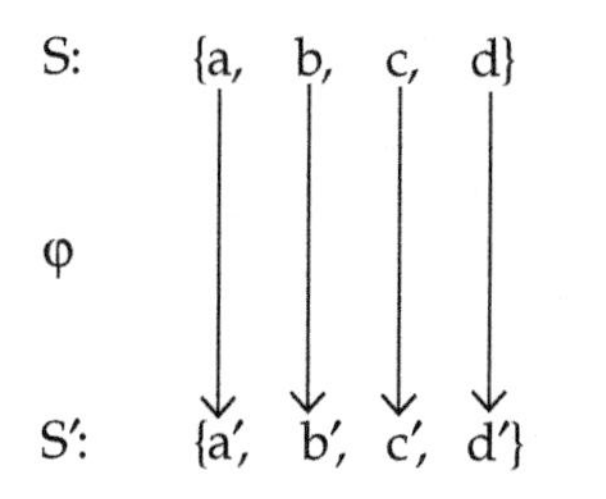

Second example:

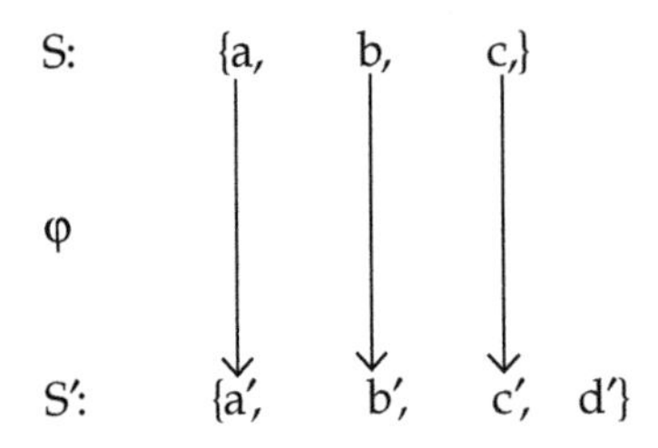

Hence, many-to-one or one-to-one mapping of the set S to the set S′ can be a mapping of S into or onto S′.

Let us examine two more examples of mapping:

1. Consider the following two sets:

$$S = \{7, 3, 4\}$$
$$\downarrow \searrow \downarrow$$
$$S' = \{1, -1, i\}$$

   The correspondence indicated by arrows is many-to-one and into mapping of S to S′ as here it associates two elements of S with a unique element of S′ and one element of S′ has been left out: it is not associated with any element of S.

2. Next, consider the following two sets:

$$S = \{7, 3, 4\}$$
$$\downarrow \searrow \swarrow \downarrow$$
$$S' = \{1, 0\}$$

   The correspondence indicated by arrows is many-to-one and onto mapping of S to S′.

1.15 Homomorphism

We have seen that many-to-one or one-to-one mapping of a set S to a set S′ under a correspondence φ can be either onto or into S′. Let us now consider a many-to-one mapping φ of a group G = {a, b, c, ...} onto a group G′ = {a′, b′, c′, ...}.

The two groups may have the same or different laws of composition. For instance, we may take the first group G as the group of all integers {..., –3, –2, –1, 0, 1, 2, 3, ...} with the law of composition as addition, whereas the second group G′ may be taken as {1, –1} with multiplication as the binary operation. The correspondence φ is such that it associates zero and all even integers in G to 1 in G′ and all odd integers in G to –1 in G′. Let $\varphi(a) \equiv a'$, $\varphi(b) \equiv b'$, and $\varphi(c) \equiv c'$, ... be the elements of G′ that correspond to the elements a, b, c, ... of G under the mapping φ. The elements φ(a), φ(b), and φ(c) of G′ may or may not be equal. Suppose that when a and b are associated to $a'(\equiv\varphi(a))$ and $b'(\equiv\varphi(b))$, the product ab (=c, say) is associated to the product $a'b'(\equiv\varphi(ab))$. This fact is expressed by stating that the correspondence between the elements of G and G′ is preserved under combination. The mapping φ is then said to be a *homomorphism* of G to G′. The group G is said to be homomorphic to the group $G' \equiv \varphi(G)$. The group $G' \equiv \varphi(G)$ is known as a *homomorphic image* or homomorph of G under φ, and the group G is said to be the *covering group* of G′. Notice that, since $(a'b') = a'b'$, for homomorphism, we must have

$$\varphi(ab) = \varphi(a)\varphi(b)$$

We now describe some simple properties of homomorphism.

Theorem 1.2

Under a homomorphism of a group G to a group G′, the identity elements correspond and the inverse of any element of G is associated to the inverse of the corresponding element of G′.

PROOF

Let φ be a homomorphism of a group G to a group G′. We will first show that under the mapping φ, the identity element of G′ corresponds to the identity element of G. We proceed in the following manner. Let a be an arbitrary element of G. Then under φ the corresponding element of G′ is denoted by φ(a). This can be expressed as

$$\varphi(a) = \varphi(ea) = \varphi(e)\,\varphi(a)$$

where the relation $\varphi(ab) = \varphi(a)\varphi(b)$ has been used. The equation $\varphi(e)\varphi(a) = \varphi(a)$ shows that φ(e), which corresponds to the identity element e of G, is the identity element of G′:

$$\varphi(e) = e'$$

We next show that the inverse of any element a of G corresponds to the inverse of the corresponding element φ(a) of G′. We have

$$e' = \varphi(e) = \varphi(aa^{-1}) = \varphi(a)\varphi(a^{-1})$$

That is, the product of $\varphi(a)$ and $\varphi(a^{-1})$ is the identity element $\varphi(e) = e'$ of G'. Therefore, one is the inverse of the other:

$$\varphi(a^{-1}) = [\varphi(a)]^{-1}$$

Hence the result. ■

1.16 Kernel

Consider a group G. Let K be a subset of G. Suppose that the subset K is such that under a homomorphism φ all its elements are associated with the identity element e' of a group G'. Then K is called the *kernel* of φ.

Consider the set of all integers $G = \{\ldots, -2, -1, 0, 1, 2, \ldots\}$ forming a group under addition, and a set $G' = \{1, -1\}$ of two elements forming a group with the law of composition as ordinary multiplication. We construct the mapping φ of G onto G' by associating zero and every even integer of G with 1 of G' and every odd integer of G with −1 of G'. This mapping is a homomorphism and the kernel of homomorphism is the set of all even integers.

We will now show that the kernel K of a homomorphism φ of a group G to a group G' is a normal subgroup of G. To prove this, we proceed as follows.

Let a and b be any two elements of the kernel K. Then, by definition, for each of them the corresponding element of G' is the identity element e':

$$\varphi(a) = e', \varphi(b) = e'$$

Consequently,

$$\varphi(ab^{-1}) = \varphi(a), \varphi(b^{-1}) = \varphi(a), [\varphi(b)]^{-1} = e' e'^{-1} = e'$$

which shows that ab^{-1} also belongs to K. Thus, K is a subgroup of G.

Also for any element c of G, we have

$$\varphi(c^{-1} ac) = \varphi(c^{-1}) \varphi(a)\varphi(c) = [\varphi(c)]^{-1}, \varphi(a)\varphi(c) = e'^{-1} e' e' = e'$$

This shows that $c^{-1} a c \in K$. Since a is arbitrary, it means that $c^{-1}Kc = K$. Hence, the kernel K is a normal subgroup of G.

PROBLEM 1.18

Show that if H is a normal subgroup of G, then the group G is homomorphic to the factor group G/H.

1.17 Isomorphism

Consider two groups G and G′ of the same order. For instance, let us take

$$G = \{1, \omega, \omega^2\}$$

and

$$G' = \{I, A, B\}$$

where 1, ω, ω^2, are cube roots of unity and

$$I = \begin{bmatrix} 1 & 0 \\ 0 & 1 \end{bmatrix}, \; A = \begin{bmatrix} -1 & -1 \\ 1 & 0 \end{bmatrix}, \; B = \begin{bmatrix} 0 & 1 \\ -1 & -1 \end{bmatrix}$$

These two groups have different laws of composition. G forms a group under ordinary multiplication, while G′ is a group under matrix multiplication. In general, the two groups may have the same or different laws of composition. In this example, each element of G is associated with a unique element of G′ under a correspondence φ indicated by arrows:

$$\begin{aligned} &G \quad G' \\ &1 \to I \\ &\omega \to A \\ &\omega^2 \to B \end{aligned}$$

It can be easily verified that the product of any two elements of G is associated with the product of the corresponding elements of G′, that is, that the correspondence is preserved under combination. For instance, with $\omega \to A$ and $\omega^2 \to B$, the product of ω and ω^2 (i.e., $\omega^3 \equiv 1$) goes to the product of A and B (i.e., AB = I). The mapping is then called an *isomorphism* of G to G′, and the group G is said to be isomorphic to the group G′. In general, consider two groups G and G′ of the same order. The two groups may or may not have the same law of composition. If under a mapping φ, one and only one element of G is associated with a unique element of G′ and the correspondence is preserved under combination, then G is said to be isomorphic to G′. We write $G \approx G'$ and read it as "G is isomorphic to G′". Since it is a one-to-one and onto mapping, if $G \approx G'$, then $G' \approx G$, and we usually say that the two groups are isomorphic to each other. The mapping is called an isomorphism. It is obviously a special case of homomorphism: a homomorphism that is one-to-one mapping is called an isomorphism.

Clearly, under an isomorphism the identity elements of two groups correspond and the inverses correspond to the inverses of the corresponding elements. Also the kernel of an isomorphism consists of the identity element alone.

Let G be the set of all integers, G = {..., –2, –1, 0, 1, 2 ...}, and G′ be the set of even integers, G′ = {..., –4, –2, 0, 2, 4 ...}. Each one of these sets forms a group under ordinary addition. Now consider a mapping φ of the group G onto the group G′ such that $\varphi(n) = 2n$. Of course, $\varphi(n)$ denotes the element of G′ corresponding to the element n of G and is equal to 2n. Clearly φ is a one-to-one mapping of G onto G′, in which any element n of G is associated with the element 2n of G′. Also, if $m, n \in G$, then the element $\varphi(m + n)$ of G′ corresponding to the element m + n of G is

$$\varphi(m+n) = 2(m+n) = 2m + 2n = \varphi(m) + \varphi(n)$$

This shows that the correspondence between the elements of two groups is preserved under combination so that the groups G and G′ are isomorphic. This example also shows that an infinite group can be isomorphic to its proper subgroup. Of course, this can happen only for an infinite group.

It is important to mention the correspondence between two isomorphic groups since the groups can be isomorphic under more than one correspondence. For instance, the two groups

$$G = \{1, \omega, \omega^2\}$$

and

$$G = \begin{bmatrix} 1 & 0 \\ 0 & 1 \end{bmatrix}, \begin{bmatrix} -1 & -1 \\ 1 & 0 \end{bmatrix}, \begin{bmatrix} 0 & 1 \\ -1 & -1 \end{bmatrix}$$

previously considered are isomorphic to each other under two different mappings as indicated:

$$\begin{array}{ccc} G & & G' \\ 1 & \leftrightarrow & \begin{bmatrix} 1 & 0 \\ 0 & 1 \end{bmatrix} \\ \omega & \leftrightarrow & \begin{bmatrix} -1 & -1 \\ 0 & 1 \end{bmatrix} \\ \omega^2 & \leftrightarrow & \begin{bmatrix} 0 & 1 \\ -1 & -1 \end{bmatrix} \end{array} \qquad \begin{array}{ccc} G & & G' \\ 1 & \leftrightarrow & \begin{bmatrix} 1 & 0 \\ 0 & 1 \end{bmatrix} \\ \omega & \leftrightarrow & \begin{bmatrix} 0 & 1 \\ -1 & -1 \end{bmatrix} \\ \omega^2 & \leftrightarrow & \begin{bmatrix} -1 & -1 \\ 0 & 1 \end{bmatrix} \end{array}$$

Since G and G′ are two isomorphic groups, the properties of G go over to the properties of G′ if the elements of G are replaced by the corresponding elements of G′ and the binary operation for G is changed to that for G′. This fact is expressed by stating that the isomorphic groups have the same structure; that is, two concrete examples of isomorphic groups are the realizations of the same abstract group. Therefore, it is sufficient to consider only one of different isomorphic groups. The concept of isomorphism is hence extremely useful. An isomorphic mapping of a group onto itself is called an *automorphism*.

Example of automorphism:

$$1 \leftrightarrow 1$$

$$\omega \leftrightarrow \omega^2$$

$$\omega^2 \leftrightarrow \omega$$

1.18 Direct Product of Groups

We will now define the direct product of groups and show how it can be used to construct new groups. Let $G' = \{a', b', c', \ldots\}$ and $G'' = \{a'', b'', c'', \ldots\}$ be two groups under binary operations $\cdot$ and $*$, respectively. Let us form the set of all ordered pairs of the type (p', q''), where the first element belongs to G', and the second belongs to G''. It is customary to denote this set either by $G' \times G''$ or by $G' \otimes G''$. We will show that under a law of composition defined by

$$(a', b'') \times (c', d'') = (a' \cdot c', b'' * d'')$$

the set $G' \otimes G''$ forms a group.

(i) The set $G' \otimes G''$ is closed under this law of composition because if (a', b'') and (c', d'') belong to $G' \otimes G''$, then the product of these two elements is

$$(a', b'') \times (c', d'') = (a' \cdot c', b'' * d'') = (f', g'')$$

and therefore belongs to $G' \otimes G''$.

(ii) The multiplication is associative because $\cdot$ and $*$ are associative.

(iii) The element (e', e''), where e' and e'' are, respectively, the identity elements of the two groups G' and G'', belongs to the set $G' \otimes G''$. Now, for any $(a', b'') \in G' \otimes G''$, we have

$$(a', b'') \times (e', e'') = (a' \cdot e', b'' * e'') = (a', b'')$$

Similarly,

$$(e', e'') \times (a', b'') = (a', b'')$$

This shows that (e', e'') is the identity element of $G' \otimes G''$.

(iv) The element (a'^{-1}, b''^{-1}) of the set $G' \otimes G''$ exists because $a' \varepsilon G'$ and $b'' \varepsilon G''$ and is the inverse of the element (a', b'') belonging to the same set. This is because

$$(a'^{-1}, b''^{-1}) \times (a', b'') = (a'^{-1} \cdot a', b''^{-1} * b'') = (e', e'')$$

Similarly,

$$(a', b'') \times (a'^{-1}, b''^{-1}) = (a' \cdot a'^{-1}, b'' * b''^{-1}) = (e', e'')$$

Hence, under the law of composition defined previously, the set $G' \otimes G''$ forms a group. The group $G' \otimes G''$ is called the direct *product* of groups G' and G''. Similarly, we can show that $G'' \otimes G'$ is a group under a similar law of composition. Of course, the groups $G' \otimes G''$ and $G'' \otimes G'$ are, in general, different.

If G' and G'' are *finite groups* of order m and n, respectively, then order of the group $G' \otimes G''$ or $G'' \otimes G'$ is mn. Thus, taking the direct product of two finite groups, each having more than one element, gives us the simplest method of forming a larger group.

If a group G is isomorphic to the group $G' \otimes G''$, then it is said to be the direct product group. With this definition, elements of the direct product group need not be in the form of pairs.

The direct product of more than two groups can be defined in the same manner.

We will now show that the set of elements

$$(a', e''), (b', e''), (c', e''), \ldots$$

forms a subgroup of $G' \otimes G''$.

We first note that all these pairs are members of the group $G' \otimes G''$. For (a', e'') and (b', e''), any two members of the given set, we have $(a', e'')(b', e'')^{-1} = (a', e'')(b'^{-1}, e''^{-1}) = (a'b'^{-1}, e'') = (f', e'')$, again a member of the set. Hence, the previously given set of elements is a subgroup of $G' \otimes G''$. We denote this subgroup by Γ':

$$\Gamma' = \{(a', e''), (b', e''), (c', e''), \ldots\}$$

Similarly, we can show that

$$\Gamma'' = \{(e', a''), (e', b''), (e'', c''), \ldots\}$$

is a subgroup of G.

PROBLEMS 1.19
If G' and G'' are two groups, show that the group $G' \otimes G''$ is isomorphic to the group $G'' \otimes G'$.

PROBLEMS 1.20
Show that the subgroups Γ' and Γ'' are, respectively, isomorphic to G' and G''.

1.19 Direct Product of Subgroups

Consider two groups $G' = \{a', b', c', \ldots\}$ and $G'' = \{a'', b'', c'', \ldots\}$ having the same law of composition. Suppose that

1. there is no common element between them except the identity and
2. the elements of G' commute with those of G''.

Let us form the set S of all those elements obtained by multiplying each element of G' with every element of G''. The set S is then given by

$$S = \{a'b'', c'd'', \ldots\}$$

that is, each of its elements is of the form $p'q''$ where the first factor belongs to G' and the second factor belongs to G''. The order, however, is immaterial as the elements of G' commute with those of G''. We will now show that this set forms a group under the common law of composition.

Consider two arbitrary elements $a'b''$, $c'd''$ of the set S. The product of these elements of S is

$$(a'b'')(c'd'') = a'b''c'd'' = a'c'b''d'' = p'q''$$

where the last but one step has been taken as the elements of G' commute with those of G''. This exhibits the closure property of the set S. It can be shown that the other properties of a group are also possessed by the elements of the set S, which therefore forms a group, say, G. Moreover, not only is G' a subset, but also a subgroup of G as all its elements belong to G and form a group under the same law of composition as for G. The same is true for G''. The group G formed in this manner is called the direct product of its subgroups G' and G''. We write it symbolically as

$$G = G' \otimes G''$$

The subgroups G' and G'' are said to be direct factors of G. If a group possesses more than two subgroups, this analysis can be easily extended to their direct product.

PROBLEMS 1.21

By using (1) and (2) listed at the outset of Section 1.19, show the following:

a. Every element g of G can be expressed in a unique manner as the product of $g' \in G'$ and $g'' \in G''$.
b. G′ and G″ are normal subgroups of G.

What will the situation be when one of the subgroups consists of the identity element only?

Example 1.2

PROBLEM

Show that the cyclic group G of order 6 $\{a, a^2, a^3, a^4, a^5, a^6 = e\}$ with the law of composition as ordinary multiplication is the direct product of its subgroups $G' = \{e, a^2, a^4\}$ and $G'' = \{e, a^3\}$.

SOLUTION

We show that the group G with the law of composition as ordinary multiplication is the direct product of its subgroups G′ and G″ (1) that do not have any common element except the identity and (2) whose elements commute with each other. We form the direct product of G′ and G″ by multiplying every element of G′ with each element of G″. We get the following set of elements:

$$\{a, a^2, a^3, a^4, a^5, a^6 = e\}$$

These elements obviously constitute the group G.

2

Group Representations

Physicists have always found it more convenient to deal with the matrix representations of a group rather than the abstract group itself. If the matrix representation of a group is reducible, then one or more invariant subspaces of it must exist with respect to the group. These invariant subspaces may be reducible or irreducible. The importance of group representations stems from the fact that various irreducible invariant subspaces of a group can be used to accommodate elementary particles with specific characteristics. Before we begin with the group representation theory, we summarize some of the material needed for it. In this context, linear vector spaces and linear operators have special significance.

2.1 Linear Vector Spaces

We first define a *field* since this term will be used in defining a linear vector space. A field F is a set, having at least two elements, in which two binary operations called addition and multiplication are defined and the set has the following properties with respect to these operations:

1. It is an Abelian group under addition.
2. Its elements other than the additive identity form an Abelian group under multiplication.
3. The multiplication is distributive over addition, that is,

$$a(b + c) = ab + ac$$

$$(a + b)c = ac + bc$$

for all a, b, c belonging to the set.

The elements of the field F are usually referred to as *scalars,* which may be real or complex.

PROBLEM 2.1

Show that the set of real numbers forms a field.

PROBLEM 2.2

Show that the set of complex numbers forms a field.

Let us now define a *linear* vector space. A set L of elements x, y, z, ..., usually called **vectors**, but not necessarily vectors in the ordinary sense, is said to define a *linear vector space* or simply a *vector space* over a field F if the following two conditions are satisfied:

1. A binary operation called *addition* or *vector addition* is defined (in general different from the addition defined for the field but we represent it by the same symbol, +, such that the set L forms an Abelian group under this binary operation).
2. Another binary operation, called *scalar multiplication* (it is different from the familiar scalar multiplication, i.e., the dot product of vectors), is defined and combines the elements of the field F and the elements of the set L such that
   (i) if $a \in F$ and $x \in L$, then $ax \in L$
   (ii) $a(x + y) = ax + ay$, for all $a \in F$ and $x, y \in L$
   (iii) $(a + b)x = ax + bx$, for all $a, b \in F$ and $x \in L$
   (iv) $(ab)x = a(bx)$, for all $a, b \in F$ and $x \in L$
   (v) $1x = x$

Note that for the equation $0.x = 0$, the zero on the left is a scalar whereas the zero on the right is an element of the vector space. However, it is conventional to use the same symbol for the both.

If the scalars are complex numbers, the space is called a *complex vector space;* if the scalars are real numbers, the space is called a *real vector space.*

To sum up, a set of elements that forms an Abelian group under addition and can be multiplied by scalars belonging to a field to give members of the set is said to form a linear vector space or vector space or linear space over that field.

The three-dimensional space of position vectors over a field of real numbers is a familiar example of a real vector space. The set of all $n \times n$ matrices with elements belonging to a field F is a linear vector space over F.

A subset M of a linear vector space L over a field F is said to be a *subspace* of L if it forms a vector space over F under the same binary operations under which L is a vector space. Clearly, M is a subgroup of L.

It may be emphasized that in a linear vector space we do not define the product of the elements of the space as a necessity. The introduction of such a product adds to the structure of the vector space. We shall define the product of the elements of a vector space later.

2.2 Linearly Independent Vectors

Consider a set of vectors

$$u_1, u_2, \ldots$$

in a linear vector space L and over a field F of scalars $a_1, a_2, \ldots$. A vector of the form

$$a_1u_1 + a_2u_2 + \cdots$$

is called a *linear combination* of the vectors $u_1, u_2, \ldots$ over F. The n vectors $u_1, u_2, \ldots, u_n$ are said to be *linearly independent* if it is impossible to make their linear combination

$$a_1u_1 + a_2u_2 + \cdots + a_nu_n$$

equal to zero except for $a_1 = a_2 = \cdots = a_n = 0$. Otherwise, the vectors are said to be linearly dependent. The maximum number of linearly independent vectors in a vector space is said to be *dimensionality* or *dimension* of the vector space. If there is a finite number of these vectors, the space is said to be of finite dimensions; otherwise, it is an infinite-dimensional vector space. A vector space is said to be spanned by the set of maximum number of linearly independent vectors in that space.

Since the maximum number of linearly independent vectors in an n-dimensional vector space is n, more than n vectors in such a space are always linearly dependent. In a two-dimensional space, two (or more) collinear vectors are always linearly dependent. If two vectors are not collinear, they are linearly independent.

2.3 Basis Vectors

Any set of n linearly independent vectors in a vector space L of dimension n is said to be a *basis* of L. It is also referred to as a set of *basis vectors* or a *coordinate system* or a *basis of representation*. We will now show that any vector in a vector space L can be expressed as a linear combination of basis vectors in that space. Let $\{u_1, u_2, \ldots, u_n\}$ be a basis of L over a field F. Let x be an arbitrary vector in L. Since n + 1 vectors $x, u_1, u_2, \ldots, u_n$ in the space L are linearly dependent, we may write

$$\beta x + b_1u_1 + b_2u_2 + \cdots + b_nu_n = 0 \tag{2.1}$$

with at least one of the scalars different from zero. However, β cannot be zero because otherwise Equation (2.1) would yield

$$b_1u_1 + b_2u_2 + \cdots + b_nu_n = 0$$

with at least one nonvanishing b_i, which contradicts the assumption that the vectors u_i are linearly independent. Hence, dividing Equation (2.1) by β and rearranging the terms, we get

$$x = -\frac{b_1}{\beta}u_1 - \frac{b_2}{\beta}u_2 - \cdots - \frac{b_n}{\beta}u_n$$

or

$$x = x_1u_1 + x_2u_2 + \cdots + x_nu_n$$

where $x_i = -b_i/\beta$. This is the desired result. The scalars $x_1, x_2, \ldots, x_n$ are called the coordinates of the vector x in the basis (or the coordinate system) $\{u_1, u_2, \ldots, u_n\}$. It may be remarked that the set of basis vectors in a vector space is not unique. We can select the sets of basis vectors in infinitely many ways. The values of the coordinates of a vector will depend on the choice of the set of basis vectors, although the vector itself will have an intrinsic significance.

2.4 Operators

Consider a linear vector space L. Let us map this vector space onto itself, that is, associate each vector x in L with a unique vector y also in L. This means that in this case mapping may be regarded as an operator that changes $x \in L$ to $y \in L$. We may therefore express it by writing

$$Ax = y$$

If the mapping is one-to-one, there exists the inverse mapping A^{-1} such that

$$A^{-1}y = x$$

An operator is specified if we know its effect on all the elements of L.

If an operator satisfies the following additional postulates:

$$A(x + y) = Ax + Ay$$

$$A(ax) = aAx$$

for all $x, y \in L$ and for all $a \in F$, then it is called a *linear operator.* A linear operator I such that $Ix = x$ for all $x \in L$ is called the *identity operator.* A linear operator A is said to be *regular* or *invertible* or *nonsingular* if there exists a linear operator A^{-1} such that $AA^{-1} = A^{-1}A = I$. If $y = Ax$, then $A^{-1}y = A^{-1}A\,x = I\,x = x$. In further discussion, we will be mainly concerned with linear operators.

An operator is said to be *antilinear* if it satisfies the following postulates:

$$A(x + y) = Ax + Ay$$

$$A(ax) = a{*}Ax$$

for all $x, y \in L$ and for all $a \in F$.

Two operators A and B are *equal* if $Ax = Bx$ for all $x \in L$. If A and B are two linear operators on a linear vector space L over a field F, then their product AB is defined to be the operator that has the same effect as that of successive applications of B and A in the same order, that is,

$$(AB)x = A(Bx) \text{ for all } x \in L$$

2.5 Unitary and Hilbert Vector Spaces

Consider a linear vector space L. Suppose that with every ordered pair of vectors x and y belonging to L, there is associated, by definition, a unique number, called scalar, belonging to a field F and denoted by (x, y). This number is called the *scalar product* or the *inner product* of x and y. The vector space L is then called a *scalar product space* or a *unitary vector space* provided the scalar product (x, y) satisfies the following conditions:

(i) $(x, y)^* = (y, x)$
(ii) $(x, x) \geq 0$, and $(x, x) = 0$ if and only if $x = 0$
(iii) $(ax + by, z) = a(x, z) + b(y, z)$ for all $a, b \in F$ and $x, y, z \in L$

where the asterisk denotes the complex conjugate of a scalar. Thus, a vector space in which a scalar product is defined is called a unitary vector space. It is clear from this list that the scalar product (x, x) is a real number. If $x \neq 0$, the scalar product (x, x) is a positive real number. The positive square root of this number, $+\sqrt{(x,x)}$, is called the *modulus* or the *norm* of the vector x and is denoted by $|x|$. Thus we may write $(x, x) = |x|^2$. If $(x, x) = 1$, then x is said to be *normalized to unity* or merely *normalized.* Two vectors x and y are said to be *orthogonal* if the scalar product $(x, y) = 0$. The vectors $k_1, k_2, \ldots, k_n$ belonging to a unitary vector space are said to form an orthonormal set if

$$(k_i, k_j) = \delta_{ij} = 1 \text{ for } i = j$$
$$= 0, \text{ otherwise}$$

To define a Hilbert space, we note that a sequence $c_1, c_2, \ldots, c_n, \ldots$ is said to be a Cauchy sequence if for every real positive number ε, however small, a finite positive integer N can be found such that for any two integers $m > N$ and $n > N$, $|c_n - c_m| < \varepsilon$.

A scalar product vector space is said to be *complete* if every Cauchy sequence of elements belonging to the space has a limit that also belongs to the space. A *complete unitary vector space* is called a *Hilbert space*. The unitary spaces of finite dimensions are necessarily complete and hence are Hilbert spaces. A space of square integrable functions in which the scalar product of two arbitrary functions is defined as

$$(\varphi, \psi) = \int \varphi^* \psi dx$$

is a Hilbert space of infinite dimensions. We will now give an example.

Let $x = (x_1, x_2, \ldots, x_n)$ and $y = (y_1, y_2, \ldots, y_n)$ be two arbitrary elements of a unitary vector space L of n dimensions. Suppose that the scalar product of the elements x and y is defined by

$$(x, y) = x_1 y_1^* + x_2 y_2^* + \cdots + x_n y_n^*$$

Then

$$|x| = \sqrt{(x,x)} = \sqrt{|x_1|^2 + |x_2|^2 + \cdots + |x_n|^2}$$

is the norm of x in L, and the vector space is a normed linear space that is complete because every convergent sequence of vectors converges to a vector within the space. This space is an n-dimensional Hilbert space. If the underlying field is real, then an n-dimensional unitary space is called an n-dimensional Euclidean space. The norm of the Euclidean space is defined by

$$|x| = \sqrt{(x,x)} = \sqrt{x_1^2 + x_2^2 + \cdots + x_n^2}$$

2.6 Matrix Representative of a Linear Operator

We will now show how a linear operator can be represented by a matrix. Consider a linear operator A that acts on any vector x in a linear vector space L of dimension n over a field F. By the very definition of an operator, it produces some vector y in the same space, that is,

$$Ax = y \in L \tag{2.2}$$

Let us choose an arbitrary basis $\{u_1, u_2, \ldots, u_n\}$ in L. Then the vectors x and y may be written as linear combinations of basis vectors:

$$x = x_1u_1 + x_2u_2 + \cdots + x_nu_n = x_iu_i \tag{2.3a}$$

$$y = y_1u_1 + y_2u_2 + \cdots + y_nu_n = y_ju_j \tag{2.3b}$$

where $x_i, y_j \in F$, $i, j = 1, 2, \ldots, n$ and, while writing x_iu_i and y_ju_j, the *summation convention* has been used. Substituting the expressions for x and y from equations (2.3) in Equation (2.2), we get

$$A\,(x_iu_i) = y_ju_j$$

or

$$y_ju_j = x_i\,(Au_i) \tag{2.4}$$

However, as $u_i \in L$, by the very definition of an operator, the vector Au_i should also belong to L, and we can therefore express it as a linear combination of the set of basis vectors $u_1, u_2, \ldots, u_n$. We may consequently write

$$\begin{aligned} Au_i = v_i &= A_{1i}u_1 + A_{2i}u_2 + \cdots + A_{ni}u_n \\ &= A_{ji}u_j, \quad i = 1, 2, \ldots, n \end{aligned} \tag{2.5}$$

It may be that $A_{1i}, A_{2i}, \ldots, A_{ni}$ are the coordinates of Au_i in the basis $u_1, u_2, \ldots, u_n$.
We have used double subscripts for the scalars A_{ji} so that the first subscript characterizes the u whose coefficient it is and the second subscript characterizes the u on which the operator A is acting. Substituting the expression for Au_i from Equation (2.5) in Equation (2.4), we obtain

$$(y_j - x_i\,A_{ji})u_j = 0 \tag{2.6}$$

Since n basis vectors u_j are linearly independent, Relation (2.6), expressing the fact that the linear combination of linearly independent vectors u_j, $j = 1, 2, \ldots, n$, is zero, can hold only if the coefficient of each u_j is zero, that is, if

$$y_j = x_iA_{ji}, \quad j = 1, 2, \ldots, n$$

Interchanging i and j, we get

$$y_i = A_{ij}x_j i = 1, 2, \ldots, n$$

Writing this set of equations in full, we get

$$\begin{aligned} y_1 &= A_{11}x_1 + A_{12}x_2 + \cdots + A_{1n}x_n \\ y_2 &= A_{21}x_1 + A_{22}x_2 + \cdots + A_{2n}x_n \\ &\cdots\cdots\cdots\cdots\cdots\cdots \\ y_n &= A_{n1}x_1 + A_{n2}x_2 + \cdots + A_{nn}x_n \end{aligned} \tag{2.7}$$

Let us define two column vectors X and Y and a square matrix **A** by the following relations:

$$X = \begin{bmatrix} x_1 \\ x_2 \\ \vdots \\ x_n \end{bmatrix}, \quad Y = \begin{bmatrix} y_1 \\ y_2 \\ \vdots \\ y_n \end{bmatrix}$$

$$\mathbf{A} = \begin{bmatrix} A_{11} & A_{12} & \cdots & A_{1n} \\ A_{21} & A_{22} & \cdots & A_{2n} \\ \vdots & \vdots & \vdots & \vdots \\ A_{n1} & A_{n2} & \cdots & A_{nn} \end{bmatrix} \tag{2.8}$$

Then the set of Equations (2.7) can be written as

$$\mathbf{A}X = Y \tag{2.9}$$

Comparing Equations (2.2) and (2.9), we notice that the knowledge of the matrix **A** with $u_1, u_2, \ldots, u_n$ as the basis of representation is equivalent to the knowledge of the operator A. The matrix **A** is called the *matrix representative* or merely *representative* of the linear operator A in the basis u_i.

It may be noticed that the order of the square matrix **A** is n, that is, the same as the dimension of the vector space in which the operator has been defined. Thus, by choosing vector spaces of different dimensions, different matrix representatives of the same operator can be obtained. Of course, these matrix representatives will be of different orders.

If B is a linear operator such that $By = z$, then $BAx = By = z$. It can be verified that BA is represented by the product of the matrices corresponding to A and B.

Let us illustrate it with an example.

Consider an operator A defined in a two-dimensional linear vector space such that

$$A\begin{bmatrix} x_1 \\ x_2 \end{bmatrix} = \begin{bmatrix} x_1 + x_2 \\ -2x_1 + 4x_2 \end{bmatrix} \tag{2.10}$$

Let us determine the corresponding matrix **A** with the basis vectors u_1 and u_2 given by

$$u_1 = \begin{bmatrix} 1 \\ 1 \end{bmatrix}, \quad u_2 = \begin{bmatrix} 1 \\ 2 \end{bmatrix}$$

Notice that the basis vectors are linearly independent: their linear combination $a_1u_1 + a_2u_2$ cannot be made zero except for $a_1 = a_2 = 0$. This is easily checked by observing that one column vector is not a constant multiple of the other. We have seen that the operator A acting on any one of the basis vectors u_1 and u_2 will yield a linear combination of these vectors:

$$Au_i = A_{ji}u_j, \quad i, \quad j = 1, 2$$

Therefore, for the basic vectors u_1 and u_2, we have

$$Au_1 = A_{11}u_1 + A_{21}u_2 \tag{2.11a}$$

$$Au_2 = A_{12}u_1 + A_{22}u_2 \tag{2.11b}$$

Now, by virtue of the given Relation (2.10) that characterizes the operator A, we have

$$Au_1 = A\begin{bmatrix} 1 \\ 1 \end{bmatrix} = \begin{bmatrix} 2 \\ 2 \end{bmatrix} = 2u_1$$

Similarly:

$$Au_2 = A\begin{bmatrix} 1 \\ 2 \end{bmatrix} = \begin{bmatrix} 3 \\ 6 \end{bmatrix} = 3u_2$$

Substituting the expressions for Au_1 and Au_2 in Equations (2.11), we get

$$(A_{11} - 2)u_1 + A_{21}u_2 = 0$$

and

$$A_{12}u_1 + (A_{22} - 3)u_2 = 0$$

Since u_1 and u_2 are linearly independent, the coefficient of each one of them in each of these two equations must be separately zero. This yields

$$A_{11} = 2,\ A_{21} = 0,\ A_{12} = 0,\ A_{22} = 3$$

The corresponding matrix **A** is

$$\mathbf{A} = \begin{bmatrix} A_{11} & A_{12} \\ A_{21} & A_{22} \end{bmatrix} = \begin{bmatrix} 2 & 0 \\ 0 & 3 \end{bmatrix}$$

2.7 Change of Basis and Matrix Representative of a Linear Operator

Let us next find out how the matrix representative of a linear operator on a vector space is transformed when a basis of representation in the vector space is changed. Consider a linear operator A acting on the vector space L over a field F such that

$$Ax = y, \quad x, y \in L \tag{2.12}$$

If $\{u_1, u_2, \ldots, u_n\}$ is a basis in the vector space L, then with respect to this basis, the vector x may be written as

$$x = x_i u_i \tag{2.13a}$$

Suppose that we change the basis of representation and take a new set of vectors $\{u'_1, u'_2, \ldots, u'_n\}$ as forming a new coordinate system. Then the same vector x may be expressed as a linear combination of primed basis vectors:

$$x = x'_j u'_j \tag{2.13b}$$

Comparing Equations (2.13a) and (2.13b), we get

$$x_i u_i = x'_j u'_j \tag{2.14}$$

Moreover, as the vectors u'_j belong to the vector space L, each one of them can be expressed as a linear combination of unprimed basis vectors u_i:

$$u'_j = c_{ji} u_i, \quad j = 1, 2, \ldots, n \tag{2.15}$$

where c_{ji} are scalars. Equation (2.15) gives u'_j in terms of the vectors u_i. We may write these equations in full as

$$u_1' = c_{1i}u_i = c_{11}u_1 + c_{12}u_2 + \cdots + c_{1n}u_n$$

$$u_2' = c_{2i}u_i = c_{21}u_1 + c_{22}u_2 + \cdots + c_{2n}u_n$$

....

$$u_n' = c_{ni}u_i = c_{n1}u_1 + c_{n2}u_2 + \cdots + c_{nn}u_n$$

Since we could take any basis as the starting point, the vectors u_i must be uniquely expressible in terms of the vectors u'_j. This is possible only if the determinant formed by the coefficients c_{ij}'s is nonzero:

$$|C| \neq 0 \tag{2.16}$$

where

$$C = \begin{bmatrix} C_{11} & C_{12} & \cdots & C_{1n} \\ C_{21} & C_{22} & \cdots & C_{2n} \\ \vdots & \vdots & \vdots & \vdots \\ C_{n1} & C_{n2} & \cdots & C_{nn} \end{bmatrix}$$

As the matrix C connecting the two bases of representation is nonsingular, its inverse exists. Now substituting the expression for u'_j from Equation (2.15) in Equation (2.14), we obtain

$$x_i u_i = x'_j (c_{ji} u_i)$$

or

$$(x_i - c_{ji} x'_j) u_i = 0$$

where c_{ji} are the elements of the matrix $\tilde{C}$, the transpose of C. Since the vectors u_i are linearly independent, the coefficient of each u_i must be separately zero. This gives

$$x_i = c_{ji} x'_j, \quad \text{where } |\tilde{C}| \neq 0$$

We have taken $|\tilde{C}| \neq 0$ because the transposition of a matrix does not change its determinant. In matrix notation, the previous relation may be written as

$$X = \tilde{C}X', \ |\tilde{C}| \neq 0 \tag{2.17}$$

where

$$X = \begin{bmatrix} x_1 \\ x_2 \\ \vdots \\ x_n \end{bmatrix}, \ X = \begin{bmatrix} x_1' \\ x_2' \\ \vdots \\ x_n' \end{bmatrix}$$

Similarly, we can show that

$$Y = \tilde{C}Y', \ |\tilde{C}| \neq 0 \tag{2.18}$$

Now if $\mathbf{A}$ and $\mathbf{A}'$ are the matrix representatives of the operator A in the unprimed and primed coordinate systems respectively, then we may write

$$Y = \mathbf{A}X \tag{2.19}$$

$$Y' = \mathbf{A}'X' \tag{2.20}$$

Substituting the expression for Y from Equation (2.18) in Equation (2.19) and simplifying, we get

$$\tilde{C}Y' = \mathbf{A}X$$

or

$$Y' = \tilde{C}^{-1}\mathbf{A}X \tag{2.21}$$

Because the matrix $\tilde{C}$ is nonsingular. Eliminating X between Equations (2.17) and (2.21), we obtain

$$Y' = \tilde{C}^{-1}\mathbf{A}\tilde{C}X' \tag{2.22}$$

Comparing Equations (2.20) and (2.22), we get

$$\mathbf{A}' = \tilde{C}^{-1}\mathbf{A}\tilde{C} \tag{2.23}$$

This equation shows that if the basis of representation in a vector space is changed, then the matrix representative $\mathbf{A}$ of a linear operator is also changed; it is actually transformed to an equivalent matrix $\mathbf{A}'$ through a nonsingular matrix $\tilde{C}$ that connects the two bases of representation. The matrices $\mathbf{A}$ and $\mathbf{A}'$ are said to be connected by a *similarity transformation* and are called *equivalent matrices.*

We have shown that by choosing vector spaces of different dimensions, we can find different matrix representatives of the same operator, the order of the representative square matrix being the same as the dimension of the space. Now we show that *a* different matrix representative of the same operator can be obtained by changing the basis of representation in the same vector space. Since there is no limit to the choice of the basis of representation, we can find in the same space an infinite number of matrix representatives of the same operator. We will illustrate it as shown below.

Suppose that in the last example, the basis vectors u_1 and u_2 are changed to

$$u_1 = \begin{bmatrix} 1 \\ 0 \end{bmatrix}, \; u_2 = \begin{bmatrix} 0 \\ 1 \end{bmatrix}$$

The new basis vectors are again linearly independent; one is not a constant multiple of the other. Then Equation (2.10), defining the operator A, gives

$$Au_1 = A\begin{bmatrix} 1 \\ 0 \end{bmatrix} = \begin{bmatrix} 1 \\ -2 \end{bmatrix}$$

Similarly, we have

$$Au_2 = A\begin{bmatrix} 0 \\ 1 \end{bmatrix} = \begin{bmatrix} 1 \\ 4 \end{bmatrix}$$

but

$$Au_i = A_{ji}u_j$$

gives

$$Au_1 = A_{11}u_1 + A_{21}u_2$$

and

$$Au_2 = A_{12}u_1 + A_{22}u_2$$

Substituting the expressions for Au_1, Au_2, u_1, and u_2 in the these equations and comparing the corresponding elements on the two sides of both the equations, we get

$$A_{11} = 1, A_{21} = -2, A_{12} = 1, A_{22} = 4$$

Hence the corresponding matrix **A** is given by

$$A = \begin{bmatrix} A_{11} & A_{12} \\ A_{21} & A_{22} \end{bmatrix} = \begin{bmatrix} 1 & 1 \\ -2 & 4 \end{bmatrix}$$

Thus, a change in the basis of representation changes the matrix representative of an operator.

Of course, matrix representative of an operator in a vector space is not unique; it depends upon the choice of the basis of representation in that space.

PROBLEM 2.3
Show that the two matrix representatives

$$\begin{bmatrix} 2 & 0 \\ 0 & 3 \end{bmatrix}$$

and

$$\begin{bmatrix} 1 & 1 \\ -2 & 4 \end{bmatrix}$$

of the operator A are connected by a *similarity* transformation. (In further discussion, we shall denote any linear operator and its matrix representative by the same letter.)

We will now consider the theory of group representations.

2.8 Group Representations

Consider an arbitrary group G = {R, S, ...}. Suppose that we can map G homomorphically onto a group of operators A, B, ... in a linear vector space L of dimension n. Then this group of operators is called a *representation* of G in the *representation space* L. We denote this group of operators by D(G) = {A, B, ...}. Frequently, the unique operator corresponding to an element R of G is denoted by D(R). Then the representation of the group G = {R, S, ...} may be written as D(G) = {D(R), D(S), ...}. The representation D(G) is said to be linear if the operators D(R), D(S), ... are linear. In further discussion, we shall restrict ourselves to linear representations. The dimension n of the vector space L is said to be the *dimension* or *degree* of the representation D(G). A group can have representations of both finite and infinite dimensions. We shall, however, confine our study to representations of finite dimensions. If G is isomorphic to D(G), then the representation D(G) is said to be *faithful*. For instance, the group {–1, 1} under ordinary multiplication is isomorphic to the group of operators $\{A, A^2 = I\}$, then the representation is faithful. On the other hand, if every element of a group is mapped onto the identity, the product is preserved under combination but the mapping is not an isomorphism, and hence the representation is not faithful.

Let us choose any basis of representation in an n-dimensional space L. Then, we can determine the matrix representatives of all the linear operators constituting the group D(G) and on which the group G is mapped homomorphically. We thus obtain a homomorphic mapping of the group G onto the group of $n \times n$ matrices. Such a representation is called a *matrix representation* of G and is also denoted by D(G). We conclude that every group G can be represented by a group of matrices D(G). If we change the basis of representation in L, we obtain another matrix representation of the same group G.

If a representation is faithful, the order of the group of matrices D(G) is the same as the order of the group G. If we choose a vector space of different dimensions, we again get a different matrix representation, the order of each matrix in this representation equal to the dimensions of the vector space. Hence, we can have different matrix representations of the same group either by choosing different sets of basis vectors in a vector space L or by considering vector spaces of different dimensions. Of course, the order of a matrix in any matrix representation will be equal to the dimension of the representation space. We distinguish among different representations of the same group by using superscripts. For example, $D^{(\alpha)}(G)$, $D^{(\beta)}(G)$, … are used to denote different representations of the group G. The dimension of the α-representation will be denoted by n_α and so on.

We will now obtain a result to be frequently used in further discussion. Suppose that a group G has been mapped homomorphically onto a set of matrices D(G) so that the correspondence between the elements of G and D(G) is preserved under combination. We will show that the set D(G) of matrices forms a group. We have

$$
\begin{array}{ll}
 & G = \{R, S, \ldots\} \\
\varphi: & \quad \searrow \; \searrow \\
 & D(G) = \{D(R), D(S), \ldots\}
\end{array}
$$

The correspondence φ is indicated by arrows. Since the correspondence between the elements of the group G and the set D(G) is preserved under combination, we have

$$(RS) \rightarrow D(RS)$$

Now

$$(RS) = RS$$

Therefore,

$$D(RS) = D(R)D(S)$$

As shown next, this equation for the closure property leads to the conclusion that D(G) forms a group.

Let us denote the identity element of G by E. For S = E, the previously given matrix equation yields

$$D(RE) = D(R)D(E)$$

or

$$D(R) = D(R)D(E)$$

That is, D(E), the element of the set D(G) corresponding to the identity element E of the group G, serves as the identity element of D(G).

For $S = R^{-1}$, equation D(RS) = D(R)D(S) yields

$$D(RR^{-1}) = D(R)D(R^{-1})$$

or

$$D(E) = D(R)D(R^{-1})$$

This shows that D(R) and $D(R^{-1})$ are inverses of each other:

$$D(R^{-1}) = [D(R)]^{-1}$$

That is, the inverse of an element D(R) of the group D(G) is the element $D(R^{-1})$ of the group D(G) corresponding to the element R^{-1} of G.

For matrices, the multiplication is always associative. Hence,

$$D(G) = \{D(R), D(S), D(T), \ldots\}$$

forms a group of matrices. In further analysis, whenever in such a situation we find that D(RS) = D(R)D(S), we would immediately conclude that D(G) forms a group of matrices.

2.9 Equivalent and Unitary Representations

The matrix representations of a group that are connected by a similarity transformation are said to be *equivalent representations*. Thus, if D(R) and D′(R) are matrices of two equivalent representations corresponding to the same element R of a group, then these must be connected to each other through a similarity transformation

$$D'(R) = C^{-1}\,D(R)C, \quad \text{for } \textit{all } R \in G$$

That is, D(R) and D′(R) are transforms of each other through a nonsingular matrix C. But we have shown that when the basis of representation is changed, the matrices of a representation are replaced by their transforms. Hence, by a suitable change of basis, equivalent representations can be obtained from each other. In fact, equivalent representations have the same structure although the corresponding matrices in the two representations are different.

The *character* of a matrix is merely another name for its trace (i.e., the sum of its diagonal elements) and is denoted by $\chi(R)$. We will show that it is an intrinsic property of D(R)—that is, that it remains invariant under a change of basis. In other words, equivalent matrices have the same character. This is shown as follows. Using the summation convention, we have

$$\begin{aligned}
\mathrm{Tr}D'(R) &= D'_{ii}(R) \\
&= (C^{-1}D(R)C)_{ii} \\
&= C^{-1}{}_{ik}D_{k\ell}(R)C_{\ell i} \\
&= C_{\ell i}C^{-1}{}_{ik}D_{k\ell}(R) \\
&= (CC^{-1})_{\ell k}D_{k\ell}(R) \\
&= (I)_{\ell k}D_{k\ell}(R) \\
&= (ID(R))_{\ell\ell} = (D(R))_{\ell\ell} \\
&= \mathrm{Tr}D(R)
\end{aligned}$$

We conclude that the equivalent representations have the same set of characters.

A representation in which all the matrices are unitary is said to be a *unitary representation*. For finite groups, it can be proved that every representation is equivalent to a unitary representation. That is, for finite groups, any matrix representation of a group can be transformed by a suitable similarity transformation into a unitary representation: if D(G) is a matrix representation of a group G, then there exists a matrix C such that the representation C^{-1} D(G) C is unitary.

2.10 Reducible and Irreducible Representations

We will now explain reducible and irreducible representations, which play a very important role in high energy physics. Consider an arbitrary group G = {R, S, T, ...}. Let D(G) = {D(R), D(S), D(T), ...} be a matrix representation of this group in a vector space L of dimension n. Each of the matrices in D(G) will be a nonsingular square matrix of order n. Suppose that either all the matrices of D(G) are in the form

$$D(R) = \begin{bmatrix} D^{(1)}(R) & A(R) \\ O & D^{(2)}(R) \end{bmatrix}$$

where R is an arbitrary element of G, $D^{(1)}(R)$ and consequently $D^{(2)}(R)$ are square matrices, and A(R) is a block matrix or that they can be brought to this form by a single similarity transformation, that is, by a single change in the basis of representation. Then the representation is said to be *reducible*. If this is not possible—that is, no single similarity transformation can bring all the matrices of the representation to this form—then the representation is said to be *irreducible*. Notice that as each square matrix D(R) of D(G) is of order n, if we choose the square matrix $D^{(1)}(R)$ of order n_1, then

(i) $D^{(2)}(R)$ will be of order $n - n_1 = n_2$, say.
(ii) A(R) will be an $n_1 \times n_2$ rectangular matrix.
(iii) O will be an $n_2 \times n_1$ null matrix.

If A(R) = O for all R in G, then the representation is said to be *fully reducible* or *decomposable*. In this case, every matrix of D(G) can be written as

$$D(R) = \begin{bmatrix} D^{(1)}(R) & 0 \\ 0 & D^{(2)}(R) \end{bmatrix}$$

where, conventionally, we have written 0 for O.

We now state an important theorem but omit its proof. This theorem is true only for finite groups. However, it can be extended to some infinite groups—the compact groups.

Theorem

If a representation is in the reducible form

$$D(R) = \begin{bmatrix} D^{(1)}(R) & A(R) \\ 0 & D^{(2)}(R) \end{bmatrix}$$

for all R in G, then by a suitable similarity transformation it can be brought to the unitary reducible form

$$D'(R) = \begin{bmatrix} D^{(1)}(R) & 0 \\ 0 & D^{(2)}(R) \end{bmatrix}$$

for all R in G, where D′(R) is unitary. ■

2.11 Complex Conjugate and Adjoint Representations

Suppose that D(G) = {D(R), D(S), ...} is a matrix representation of a group G in a vector space. If we take the complex conjugate D*(R) of each matrix D(R) of the matrix representation D(G) of the group G, we get the *complex conjugate representation* D*(G) of the group G since

$$D^*(RS) = [D(RS)]^* = [D(R)\, D(S)]^* = D^*(R)\, D^*(S)$$

If we take the inverse and the transpose of each of the matrices of D(G), we again obtain a matrix representation, say, $\bar{D}(G)$. This is because

$$\bar{D}(RS) = \tilde{D}^{-1}(RS) = [\tilde{D}(RS)]^{-1} = [\tilde{D}(S)\tilde{D}(R)]^{-1}$$

$$= \tilde{D}^{-1}(R)\tilde{D}^{-1}(S) = \bar{D}(R)\bar{D}(S)$$

This representation is called the *adjoint representation* of the group G.

2.12 Construction of Representations by Addition

Let us consider two matrix representations $D^{(1)}(G) = \{D^{(1)}(R), D^{(1)}(S), \ldots\}$ and $D^{(2)}(G) = \{D^{(2)}(R), D^{(2)}(S), \ldots\}$ of a group G in vector spaces of dimensions n_1 and n_2, respectively (n_2 may or may not be equal to n_1). The square matrices in the two representations will be of orders n_1 and n_2, respectively. We will show in the following that we can always construct a new representation by combining two given representations.

Let us form *square* matrices of the type

$$D(R) = \begin{bmatrix} D^{(1)}(R) & 0 \\ 0 & D^{(2)}(R) \end{bmatrix}$$

Since $D^{(1)}(R)$ and $D^{(2)}(R)$ are of orders n_1 and n_2, respectively, the null matrices in the right and left corners of the previous matrix must be of orders $n_1 \times n_2$ and $n_2 \times n_1$. The matrix D(R) will be of order $n_1 + n_2$ (= n, say). We will now show that the set D(G) of matrices is a fully reducible matrix representation of the group G, but in a space of $n_1 + n_2$ dimensions. We first prove that the matrices D(G) form a group.

If D(R) and D(S) are two elements of the set D(G), then by using block multiplication for matrices, we have

$$D(R)D(S) = \begin{bmatrix} D^{(1)}(R) & 0 \\ 0 & D^{(2)}(R) \end{bmatrix} \begin{bmatrix} D^{(1)}(R) & 0 \\ 0 & D^{(2)}(R) \end{bmatrix}$$

$$= \begin{bmatrix} D^{(1)}(R)D^{(1)}(S) & 0 \\ 0 & D^{(2)}(R)D^{(2)}(S) \end{bmatrix}$$

Since $D^{(1)}(G)$ is a matrix representative of the group G, the correspondence between the elements of G and $D^{(1)}(G)$ must be preserved:

$$D^{(1)}(R)D^{(1)}(S) = D^{(1)}(RS)$$

with a similar relation for the matrices of $D^{(2)}(G)$. Therefore, we have

$$D(R)D(S) = \begin{bmatrix} D^{(1)}(RS) & 0 \\ 0 & D^{(2)}(RS) \end{bmatrix} = D(RS) \tag{2.24}$$

This shows that the set D(G) forms a group under matrix multiplication. It is clear from Equation (2.24) that the group G is mapped homomorphically on the matrix group D(G). Hence, D(G) is a matrix representation of the group G in a vector space of $(n_1 + n_2)$ dimensions.

Thus, from two matrix representations $D^{(1)}(G)$ and $D^{(2)}(G)$ of a group G in vector spaces L_1 and L_2 of dimensions n_1 and n_2, respectively, we can form another matrix representation D(G) in a vector space L of higher dimension $n_1 + n_2$. This process is called the *addition of representations*. The representation D(G) can then be added to $D^{(1)}(G)$ or $D^{(2)}(G)$ to get new representations, and this process can be continued. The addition of representations is therefore a very useful technique for the construction of new representations of higher order.

We usually express the result obtained by the addition of representations as

$$D = D^{(1)} + D^{(2)}$$

The rearrangement of terms in the sum of representations is evidently allowed.

2.13 Analysis of Representations

We will now consider the reverse process. Suppose that D(G) = {D(R), D(S), …} is a *reducible* matrix representation of a group G = {R, S, …} in a vector space of dimension n so that an arbitrary matrix D(R) of this representation is a square matrix of order n and is given by

$$D(R)=\begin{bmatrix} D^{(1)}(R) & A(R) \\ 0 & D^{(2)}(R) \end{bmatrix} \tag{2.25}$$

where the matrix $D^{(1)}(R)$ is chosen to be square and of order n_1. Then $D^{(2)}(R)$ will be a square matrix of order n_2, where $n_2 = n - n_1$.

We will prove that in this case $D^{(1)}(G) = \{D^{(1)}(R), D^{(1)}(S), \ldots\}$ and $D^{(2)}(G) = \{D^{(2)}(R), D^{(2)}(S), \ldots\}$ are also matrix representations of the group G but in vector spaces of dimensions n_1 and n_2, respectively. We proceed as follows. As D(G) is a matrix representation of the group G, as already shown, we must have

$$D(R)\,D(S) = D(RS)$$

Substituting the expressions for these matrices, we get

$$\begin{bmatrix} D^{(1)}(R) & A(R) \\ 0 & D^{(2)}(R) \end{bmatrix}\begin{bmatrix} D^{(1)}(S) & A(S) \\ 0 & D^{(2)}(S) \end{bmatrix} = \begin{bmatrix} D^{(1)}(RS) & A(RS) \\ 0 & D^{(2)}(RS) \end{bmatrix}$$

or

$$\begin{bmatrix} D^{(1)}(R)D^{(1)}(S) & D^{(1)}(R)A(S) + A(R)D^{(2)}(S) \\ 0 & D^{(2)}(R)D^{(2)}(S) \end{bmatrix} = \begin{bmatrix} D^{(1)}(RS) & A(RS) \\ 0 & D^{(2)}(RS) \end{bmatrix}$$

Comparing the corresponding blocks, we get

$$D^{(1)}(R)\,D^{(1)}(S) = D^{(1)}(RS)$$

$$D^{(2)}(R)\,D^{(2)}(S) = D^{(2)}(RS)$$

These relations show that $D^{(1)}(G)$ and $D^{(2)}(G)$ are also matrix representations of the group G. These representations may or may not be reducible.

This process can be continued. For instance, suppose that $D^{(1)}(G)$ is a reducible representation. That is, either all the matrices of $D^{(1)}(G)$ are of the form

$$D^1(R) = \begin{bmatrix} D^{(3)}(R) & A'(R) \\ 0 & D^{(4)}(R) \end{bmatrix}$$

or can be brought to this form by a single similarity transformation (i.e., by a single change of basis). Then, as analyzed already, $D^{(3)}(G)$ and $D^{(4)}(G)$ will also be matrix representations of the group G. If $D^{(3)}(G)$ is in a space of dimension n_3, then as $D^{(1)}(G)$ is of dimension n_1, $D^{(4)}(G)$ must be in a space of dimension $n_1 - n_3$, say, n_4. The same procedure can be applied to $D^{(2)}(R)$. We can continue that way until the process comes to an end and we get only irreducible representations (IRs) of the group G. Of course, the IRs cannot be decomposed into block-diagonal form of the aforementioned type.

2.14 Irreducible Invariant Subspace

If under the action of a group G the vectors of a subspace M of a linear vector space L are transformed among themselves, the subspace M is said to be *invariant* under G. If an *invariant subspace* does *not contain* any smaller invariant subspace, it is said to be an *irreducible invariant subspace* or a *multiplet* of the vector space L.

2.15 Matrix Representations and Invariant Subspaces

We will now show with the help of an example that if a group G has a reducible matrix representation in a vector space L, then, with respect to G, the space L contains invariant subspaces. These invariant subspaces may or may not be irreducible.

Consider a reducible matrix representation D(G) of a group G in a three-dimensional linear vector space L over a field F. Then each of its matrices can be put in the form

$$\begin{bmatrix} a & b & c \\ d & e & f \\ 0 & 0 & g \end{bmatrix}$$

where a, b, c, d, e, f, g $\in$ F. Let L_{12} be a two-dimensional subspace of the vector space L. This subspace consists of all vectors that have their third component equal to zero so that any vector of L_{12} may be written as

$$\begin{bmatrix} x_1 \\ x_2 \\ 0 \end{bmatrix}$$

Then, operating on this vector by the previous matrix, we get

$$\begin{bmatrix} a & b & c \\ d & e & f \\ 0 & 0 & g \end{bmatrix} \begin{bmatrix} x_1 \\ x_2 \\ 0 \end{bmatrix} = \begin{bmatrix} ax_1 + bx_2 \\ cx_1 + dx_2 \\ 0 \end{bmatrix}$$

This shows that under D(G) the vectors of the subspace L_{12}, each having at least its third component equal to zero, are transformed among themselves. Hence, this subspace is invariant with respect to the group of matrix representations D(G).

Let us next consider the one-dimensional subspace L_3 of L with vectors of the type

$$\begin{bmatrix} 0 \\ 0 \\ x_3 \end{bmatrix}$$

Then, we have

$$\begin{bmatrix} a & b & c \\ d & e & f \\ 0 & 0 & g \end{bmatrix} \begin{bmatrix} 0 \\ 0 \\ x_3 \end{bmatrix} = \begin{bmatrix} cx_3 \\ fx_3 \\ gx_3 \end{bmatrix}$$

The first two components of the vector thus obtained are not zero. Hence, under D(G) the vectors of L_3 are not transformed among themselves. Under the action of D(G) they leave the subspace and are transformed into those vectors of L which do not belong to L_3. That is, the subspace L_3 is not invariant under D(G).

Similarly, it can be shown that the one-dimensional subspaces L_1 and L_2 spanned by vectors of the type

$$\begin{bmatrix} x_1 \\ 0 \\ 0 \end{bmatrix}$$

and

$$\begin{bmatrix} 0 \\ x_2 \\ 0 \end{bmatrix}$$

are not invariant under D(G). The two-dimensional invariant subspace L_{12} contains only two subspaces L_1 and L_2, both of which are not invariant under G. The two-dimensional invariant subspace L_{12} does not contain any smaller invariant subspace. Hence, it is an irreducible invariant subspace.

It can also be shown that the subspaces L_{23} and L_{31} are not invariant under D(G).

We next consider a matrix representation that is fully reducible in a three-dimensional space L such that each one of its matrices can be written as

$$\begin{bmatrix} a & b & 0 \\ d & e & 0 \\ 0 & 0 & g \end{bmatrix}$$

It can be easily shown that in this case there are only two invariant subspaces L_{12} and L_3, known as *complementary subspaces*. However, L_3 is not a smaller subspace of L_{12}. Therefore, L_{12} is again an irreducible invariant subspace. The space L is said to be *decomposed into the direct sum* of L_{12} and L_3: $L = L_{12} + L_3$. The representation D(G) is usually written as

$$D = D^{(2)} + D^{(1)}$$

where $D^{(2)}$ and $D^{(1)}$ are representations in vector spaces of dimensions two and one, respectively.

Let us generalize to n dimensions. Then, as we will be dealing with a fully reducible representation of a group in a vector space of n dimensions, each matrix can be written in the form

$$D(R) = \begin{bmatrix} D^{(1)}(R) & 0 \\ 0 & D^{(2)}(R) \end{bmatrix}$$

where, if $D^{(1)}(R)$ is of order m, then $D^{(2)}(R)$ is of order n − m. Following the arguments given in the analysis of the matrix representation in a three-dimensional vector space, one can show that two invariant subspaces of L must exist: an m-dimensional subspace; and an (n – m)-dimensional, subspace with vectors of the type

$$\begin{bmatrix} x_1 \\ x_2 \\ \vdots \\ x_m \\ 0 \\ \vdots \\ 0 \end{bmatrix}$$

and

$$\begin{bmatrix} 0 \\ 0 \\ \vdots \\ x_{m+1} \\ \\ \vdots \\ x_n \end{bmatrix}$$

respectively. Since the vectors of an invariant subspace are transformed among themselves, the transformations of the group do not couple the vectors of these two invariant subspaces. We can, therefore, treat the invariant subspaces independently of one another.

We can now examine whether the representations $D^{(1)}(G)$ and $D^{(2)}(G)$ of dimensions m and n – m, respectively, are in turn reducible or not. And this process can be continued. By continuing this process, we can finally get irreducible representations of the group. We can then write

$$D = \mathbf{D}^{(1)} + \mathbf{D}^{(2)} + \cdots + \mathbf{D}^{(k)}$$

where $\mathbf{D}^{(i)}(G)$ are irreducible representations of the same group. We say that the representation D is fully reducible to the previously given sum. Several of these irreducible representations may be equivalent. Such representations are not counted as distinct, and we may use the same symbol for them

$$D = a_1\mathbf{D}^{(1)} + a_2\mathbf{D}^{(2)} + \cdots + a_r\mathbf{D}^{(r)}$$

where a_i are positive integers.

To sum up, a matrix representation is either irreducible or can be expressed as a sum of irreducible representations.

We shall now show that if a vector space has an invariant subspace under a group of transformations, then the matrix representation of the group in this space must be reducible.

Let

$$\begin{bmatrix} a_{11} & a_{12} & a_{13} \\ a_{21} & a_{22} & a_{23} \\ a_{31} & a_{32} & a_{33} \end{bmatrix} \tag{2.26}$$

be an arbitrary matrix of the matrix representation of the given group in a vector space of three dimensions. Let L_{12} be an invariant subspace of it. Then for all the vectors of this subspace, at least the third component must be zero. But

$$\begin{bmatrix} a_{11} & a_{12} & a_{13} \\ a_{21} & a_{22} & a_{23} \\ a_{31} & a_{32} & a_{33} \end{bmatrix} \begin{bmatrix} x_1 \\ x_2 \\ 0 \end{bmatrix} = \begin{bmatrix} a_{11}x_1 + a_{12}x_2 \\ a_{21}x_1 + a_{22}x_2 \\ a_{31}x_1 + a_{32}x_2 \end{bmatrix}$$

Since the column vector on the right-hand side must be a member of the subspace, its third component should be zero. For this to be true for all x_1, x_2, we must have

$$a_{31} = a_{32} = 0$$

This reduces Matrix (2.26) to the form

$$\begin{bmatrix} a_{11} & a_{12} & a_{13} \\ a_{21} & a_{22} & a_{23} \\ 0 & 0 & a_{33} \end{bmatrix}$$

This is true for every matrix of the group. Hence, the matrix representation is reducible.

It may be noticed that in this case $D^{(1)}(G) = \{D^{(1)}(R), D^{(1)}(S), \ldots\}$, where

$$D^{1}(R) = \begin{bmatrix} a_{11} & a_{12} \\ a_{21} & a_{22} \end{bmatrix}$$

and so forth will be matrix representations of the group G in the two-dimensional space L_{12}. This result can be easily generalized.

PROBLEM 2.4
Show that if a vector space L has two complementary invariant subspaces under a matrix representation D(G) of a group G, then D(G) must be fully reducible.

2.16 Product Representations

We have seen that new matrix representations of a group can be formed by the "addition" of given matrix representations of the same group. We will now show that new representations can also be constructed by taking the direct product of the given representations. For this purpose, we will first explain the meaning of the *direct product of matrices.*

Consider an m × n matrix A and a p × q matrix B. Then the matrix C is called the direct product of A and B provided its elements are obtained by multiplying *each element of A with every element of B and are arranged in a specified manner.* Symbolically, we write this as

$$C = A \times B$$

If A_{ij} and B_{kl}, respectively, denote the elements of the matrices A and B, then a convenient labeling of the elements of C is defined by

$$C_{ik,j\ell} = A_{ij}B_{k\ell} \tag{2.27}$$

We arrange the elements of C such that all the elements with the same ik value are in the same row, while all the elements with the same $j\ell$ value are in the same column. Moreover, the arrangement is in the dictionary order. We shall illustrate it with an example.

Suppose that two matrices A and B are given by

$$A = \begin{bmatrix} a_{11} & a_{12} \\ a_{21} & a_{22} \end{bmatrix}$$

$$B = \begin{bmatrix} b_{11} \\ b_{21} \\ b_{31} \end{bmatrix}$$

Then the product of each element of A with every element of B yields the following set of elements:

$$a_{11}b_{11} = c_{11,11},\ a_{11}b_{21} = c_{12,11},\ a_{11}b_{31} = c_{13,11}$$

$$a_{12}b_{11} = c_{11,21},\ a_{12}b_{21} = c_{12,21},\ a_{12}b_{31} = c_{13,11}$$

$$a_{21}b_{11} = c_{21,11},\ a_{21}b_{21} = c_{22,11},\ a_{21}b_{31} = c_{23,11}$$

$$a_{22}b_{11} = c_{21,21},\ a_{22}b_{21} = c_{22,21},\ a_{22}b_{31} = c_{23,21}$$

Arranging the element $c_{ik,j\ell}$ in accordance with the previously described rules, we obtain

$$\begin{bmatrix} c_{11,11} & c_{11,21} \\ c_{12,11} & c_{12,21} \\ c_{13,11} & c_{13,21} \\ c_{21,11} & c_{21,21} \\ c_{22,11} & c_{22,21} \\ c_{23,11} & c_{23,21} \end{bmatrix}$$

The direct product, it may be noticed, is a commutative operation because the same set of elements is obtained in whatever order all products of the elements of A and B are taken as they are arranged according to some specific rules.

It can be shown that if each distinct pair ik is used to label a row and each distinct pair $j\ell$ labels a column, and elements are arranged in the dictionary order, then the rule for the multiplication of product matrices is exactly the same as that for ordinary matrices.

We will now consider the direct product of two irreducible matrix representations (IRs) of a group G and show that it is also a matrix representation of G. Let n_α and n_β be the dimensions of two irreducible representations $D^{(\alpha)}(G)$ and $D^{(\beta)}(G)$ of G. By virtue of Relation (2.27), the elements of the matrix $D^{(\alpha\times\beta)}$, which stands for the direct product of the matrices $D^{(\alpha)}$ and $D^{(\beta)}$, are given by

$$D^{(\alpha\times\beta)}_{ik,j\ell}(R) = D^{(\alpha)}_{ij}(R) \times D^{(\beta)}_{k\ell}(R)$$

We will now show that the matrices $D^{(\alpha\times\beta)}(R)$, $D^{(\alpha\times\beta)}(S)$,... form another matrix representation of the group G. To prove this, we proceed as follows. We know that

$$D^{(\alpha\times\beta)}_{il,kj}(R) = D^{(\alpha)}_{ik}(R)\, D^{(\beta)}_{lj}(R)$$

Therefore, we have

$$D^{(\alpha\times\beta)}_{i\ell,kj}(R)\,D^{(\alpha\times\beta)}_{kj,mn}(S) = D^{(\alpha)}_{ik}(R)D^{(\beta)}_{\ell j}(R)D^{(\alpha)}_{km}(S)D^{(\beta)}_{jn}(S)$$

$$= D^{(\alpha)}_{ik}(R)D^{(\alpha)}_{km}(S)D^{(\beta)}_{\ell j}(R)D^{(\alpha)}_{jn}(S)$$

$$= D^{(\alpha)}_{im}(RS)D^{(\beta)}_{\ell j}(RS)$$

$$= D^{(\alpha\times\beta)}_{i\ell,mn}(RS)$$

This relation shows that $D^{(\alpha\times\beta)}(G)$ is another matrix representation of the group G. This new representation is called the *Kronecker* or *direct product* of the representations $D^{(\alpha)}(G)$ and $D^{(\beta)}(G)$. Symbolically

$$D^{(\alpha\times\beta)}(G) = D^{(\alpha)}(G) \times D^{(\beta)}(G)$$

Hence, once two representations are known, many others can be generated by forming their direct products.

The product representation $D^{(\alpha\times\beta)}(G)$ is in general reducible. If the group is finite or simple and compact, $D^{(\alpha\times\beta)}(G)$ can be decomposed into a direct sum of irreducible representations:

$$D^{(\alpha\times\beta)}(G) = \sum_i a_i D^{(\gamma)}_i(G)$$

where the number a_i denotes how many times the irreducible representation $D^{(\gamma)}_i$ appears in the sum. This decomposition is called the Clebsch–Gordan series. If $a_i = 0$ or 1 for all i, the product representation is said to be simply reducible.

We finally prove that the character of the direct product of two representations is equal to the product of the characters of the representations. We have

$$\chi(R) = D^{(\alpha\times\beta)}_{ij,ij}(R)$$

$$= D^{\alpha}_{ii}(R)D^{\alpha}_{jj}(R)$$

$$= \chi^{(\alpha)}(R)\chi^{(\beta)}(R)$$

This is true for all $R \in G$. Hence, the result.

We will now state an important theorem that determines whether any matrix representation of a finite group is irreducible, but we omit its proof.

Theorem

The necessary and sufficient condition for a matrix representation D(G) of a finite group G of order g to be irreducible is that its characters satisfy the relation RG.

$$\sum_{R \in G} \chi^*(R)\chi(R) = g$$

We now state and prove an important lemma.

Schur's Lemma

If a matrix commutes with all the matrices of an irreducible representation of a group, then it must be a multiple of unit matrix.

PROOF

Let D(R) be an arbitrary matrix element of an irreducible representation D(G) of a group G in a linear space L of dimension n. Then D(R) is an $n \times n$ matrix. Let A be a matrix which commutes with D(R) for all $R \in G$, that is,

$$A\,D(R) = D(R)A, \text{ for all } R \in G$$

We will prove that the matrix A of order n is a multiple of the unit matrix. In general, such a matrix has n eigenvalues and n eigenvectors. Let $x \in L$ be an eigenvector of the matrix A belonging to the eigenvalue a. Then

$$Ax = ax$$

Consider the vector D(R)x. Since A commutes with D(R), we have

$$A[D(R)x] = AD(R)x = D(R)Ax = D(R)\,ax = a[D(R)x]$$

Thus, [D(R)x] is also an eigenvector of A belonging to the same eigenvalue a. Since D(R) is an arbitrary matrix of the representation, it follows that [D(R)x], for all $R \in G$, is an eigenvector of A belonging to the same eigenvalue a. It is easy to show that any linear combination of these eigenvectors is also an eigenvector belonging to the same eigenvalue a. Thus the eigenvectors of A form a linear space L′ whose elements transform among themselves under D(R) for all $R \in G$. That is, L′ is an invariant subspace of L. Since the representation D(a) is an irreducible representation, no proper subspace of L is invariant under D(G). It therefore follows that $L' \equiv L$ and thus all the elements of L are eigenvectors of A belonging to the eigenvalue a. Hence, each of the n eigenvalues of A is equal to a. This implies that $A = aI$, where I is the $n \times n$ unit matrix. Hence, the lemma. ■

3

Continuous Groups

Continuous groups, particularly the transformation groups such as SU(3) and SU(6), have played a significant role in the development of elementary particle physics and have given rise to spectacular progress in our understanding of particles and their interactions. In this chapter, we will study continuous groups and their properties. The subject, however, will not be developed in a mathematically rigorous manner. Our aim is to introduce the concept of continuous groups and discuss their characteristics in a manner easily comprehensible to physicists.

3.1 Definition of a Continuous Group

A group that contains an infinite number of elements is called an infinite group. The set of all positive, negative, and zero integers and the set of real numbers are two examples of infinite groups under addition. In the first example, the number of elements is denumerably infinite; that is, it is countable. Such a group is said to be a *discrete group*. It never needs more than one discrete real parameter to label its elements. In the second example, the group is nondenumerably infinite; that is, it is noncountable. Such a group is called a *continuous group*. This particular continuous group is parameterized by only one continuously varying real parameter; all the group elements can be obtained by a continuous variation of the real parameter. In general, a continuous group may be characterized by one or more continuously varying (real or complex) parameters. If r is the minimum number of continuously varying real parameters required to characterize a continuous group, then the group is termed an *r-parameter continuous group* and the parameters are called *essential parameters*. If r is finite, the group is said to be a *finite continuous group* and the number r is known as the *order* of the continuous group. We shall restrict ourselves only to continuous groups for which the number of essential parameters is finite.

Compact groups constitute an important class of continuous groups. To understand the concept of a compact group, it is necessary to define two terms: *bounded* and *closed*. A set of numbers is said to be bounded if in its absolute value every number in the set is either equal to or less than a given positive number. It is said to be closed if the limit of every convergent sequence of points in the set also lies in the set. For instance, the set of real numbers between zero and one is bounded: none of its members can exceed one. It is also closed if both the points, viz., zero and one, are included in the set. Now, the parameters of a continuous group may vary continuously either in an infinite range or be confined to some finite domain. If the domain of variation of all the parameters is bounded and closed, the group is said to be *compact*.

A *mixed continuous group* is defined as a continuous group that, in addition to continuously varying parameters, requires a discrete label to characterize all its elements.

PROBLEM 3.1

Prove that any one-parameter continuous group is Abelian.

3.2 Groups of Linear Transformations

The groups whose elements can be realized by geometrical or physical transformations play a significant role in physics. These groups are called *transformation groups*. Such groups are usually defined in terms of linear transformations on a set of variables. Suppose that the members of two sets of variables (x_1, x_2) and (x_1', x_2') in a two-dimensional vector space are related to each other by the equations

$$\begin{aligned} x_1' &= a_{11}x_1 + a_{12}x_2 \\ x_2' &= a_{21}x_1 + a_{22}x_2 \end{aligned} \tag{3.1a}$$

where a_{ij} are scalars that can vary continuously. These equations are said to describe homogeneous linear transformations. If the inverse of a transformation exists, the determinant of its coefficients must be nonzero:

$$\begin{vmatrix} a_{11} & a_{12} \\ a_{21} & a_{22} \end{vmatrix} \neq 0 \tag{3.1b}$$

The transformation is then said to be nonsingular. In matrix notation, Equations (3.1a) and (3.1b) can be written as

$$x' = Ax, \qquad |A| \neq 0 \tag{3.2}$$

where

$$x' = \begin{bmatrix} x_1' \\ x_2' \end{bmatrix}, \ x = \begin{bmatrix} x_1 \\ x_2 \end{bmatrix}, \ A = \begin{bmatrix} a_{11} & a_{12} \\ a_{21} & a_{22} \end{bmatrix}$$

Here, x and x′ are column vectors having two elements each, and A is a square matrix of order 2 and is called the *transformation matrix*. The transformation is usually called a *homogeneous linear transformation* in a vector space of two dimensions. Let us examine a specific example.

Consider a rectangular Cartesian coordinate system S in a *two-dimensional real space* with O(0,0) as its origin. Let **r** be the position vector of any point P with respect to the origin O. Let (x_1, x_2) be the coordinates of the point P in S. If (x_1', x_2') are the coordinates of the same point P with respect to another frame of reference which has been obtained by rotating S about its origin through an angle φ, say, in the anticlockwise direction, and displacing it through a distance **b**, then the primed and unprimed coordinates of the same point P are related to each other by

$$\begin{aligned} x_1' &= a_{11}x_1 + a_{12}x_2 + b_1 \\ x_2' &= a_{21}x_1 + a_{22}x_2 + b_2 \end{aligned} \tag{3.3}$$

where a_{ij} are real and denote the cosines of the angles between the corresponding axes of S and S′, and $\mathbf{b} = (b_1,b_1)$. Of course, the parameters are allowed to vary continuously. Equations (3.3) are said to describe *inhomogeneous linear transformations*. If the frame S′ is obtained from the frame S merely by rotating it about its origin, then $\mathbf{b} = 0$ and Equations (3.3) reduce to

$$\begin{aligned} x_1' &= a_{11}x_1 + a_{12}x_2 \\ x_2' &= a_{21}x_1 + a_{22}x_2 \end{aligned} \tag{3.1a′}$$

These equations reflect how the coordinates of the point P or the components of the position vector **r** in the primed frame of reference are related to the coordinates of the same point P or the components of the same vector **r** in the unprimed frame of reference. Of course, one frame of reference is obtained from the other by just rotating it about its origin. This is shown in Figure 3.1A. As in this example the inverse of a transformation exists, the determinant of its coefficients must be nonzero:

$$\begin{vmatrix} a_{11} & a_{12} \\ a_{21} & a_{22} \end{vmatrix} \neq 0$$

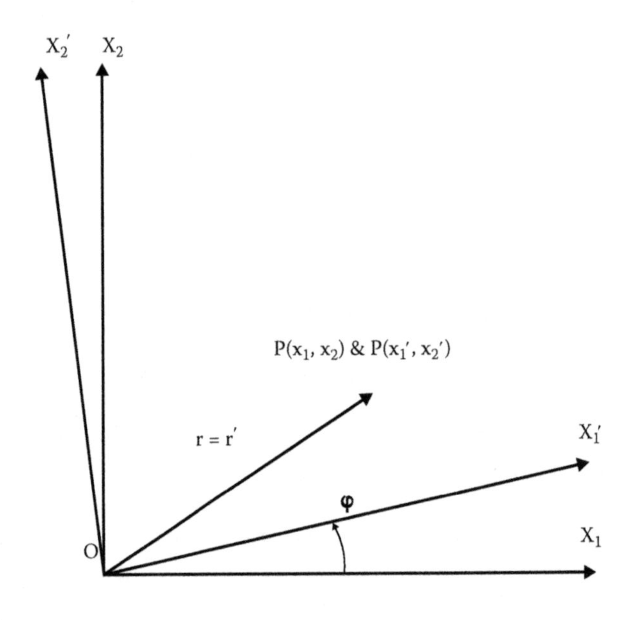

FIGURE 3.1A
Rotation of coordinate system.

The same situation can be considered with respect to a fixed coordinate system. Instead of having anticlockwise rotation of the coordinate system, we may rotate the position vector **r** clockwise by an angle φ to a new vector **r**′. This is shown in Figure 3.1B. The components of the new vector will be related to those of the old vector by the same equations—(3.1) or (3.2)—that describe the coordinate transformation. The linear transformation now maps the point with coordinates (x_1, x_2) into the point (x_1', x_2') with

$$\begin{aligned} x_1' &= a_{11}x_1 + a_{12}x_2 \\ x_2' &= a_{21}x_1 + a_{22}x_2 \end{aligned} \qquad \begin{vmatrix} a_{11} & a_{12} \\ a_{21} & a_{22} \end{vmatrix} \neq 0$$

Hence, this set of equations or the equivalent matrix equation

$$x' = Ax, \qquad |A| \neq 0$$

can be interpreted either as a rotation of the coordinate system (say, in the anticlockwise direction) or as a rotation of the vector (in the clockwise direction), keeping the coordinate system fixed. Of course, in both cases, mathematics remains the same.

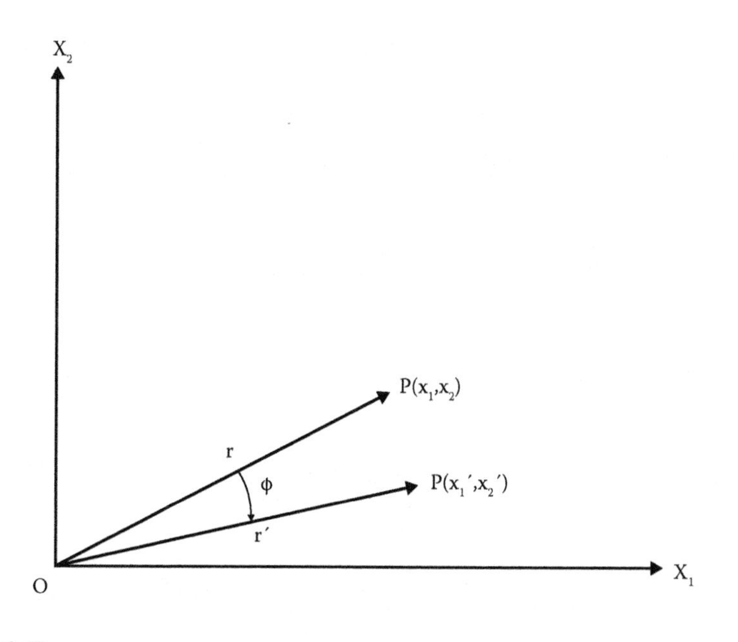

FIGURE 3.1B
Rotation of position vector in a fixed coordinate system.

The transformation T that takes x to x′ through A is usually written as

$$T: x \rightarrow x' = Ax$$

or

$$T: x' = Ax$$

We will now show that the set of all nonsingular homogeneous linear transformations in a two-dimensional real vector space forms a continuous group under successive application of transformations. Consider the set L_2 of all nonsingular homogeneous linear transformations T_1, T_2, ... in a two-dimensional real space:

$$T_1: x' = Ax, \qquad |A| \neq 0$$

$$T_2: x'' = Bx', \qquad |B| \neq 0$$

....

where each of the transformation matrices is a nonsingular real square matrix of order 2. The product T_2T_1 of the transformations T_1 and T_2 is defined to be the transformation that is obtained by applying T_1 followed by T_2; that is, first T_1 takes x to x′ and then T_2 takes x′ to x″ so that

$$T_2T_1: x'' = Bx' = BAx$$

This means that the combined operation T_2T_1 takes x to x″ through the transformation matrix BA. Thus, when the matrices A and B correspond to the transformations T_1 and T_2, respectively, the matrix BA corresponds to the transformation T_2T_1 and is itself a square matrix of order 2. Now $|BA|=|B|$ $|A| \neq 0$ so that T_2T_1 is a nonsingular transformation belonging to L_2. This is the closure property for the members of the set.

The multiplication of transformations is associative because both the operations $(T_3T_2)T_1$ and $T_3(T_2T_1)$ consist of applying T_1, T_2, and T_3 in that order.

The homogeneous linear transformation

$$x' = Ix = x, \qquad |I| \neq 0$$

is called the *identity transformation.* I is a 2 × 2 unit matrix. Evidently, this transformation keeps things unchanged; when applied to any transformation of the set, it reproduces that transformation.

If a transformation T changes the coordinates (x_1, x_2) to (x_1', x_2'), then its inverse is defined as the transformation that changes the coordinates back from (x_1', x_2') to (x_1, x_2); in other words, if T takes x to $x' = Ax$, then T^{-1} takes x′ to $x = A^{-1}x'$ so that $T^{-1}T$ takes x back to itself.

Thus, under successive application of transformations as the binary operation, the set of all nonsingular homogeneous linear transformations in a two-dimensional real space forms a group. This group is called as the *general linear group in a real space of two dimensions* and is denoted by GL(2, R). It can be shown that the set of all nonsingular homogeneous linear transformations in a two-dimensional complex space also forms a group under successive application of transformations as the binary operation. This group is known as the *general linear group in a complex space of two dimensions* and is denoted by GL(2, C) or more frequently by GL(2). In matrix form, any transformation of the group GL(2) is also given by

$$x' = Ax, \qquad |A| \neq 0$$

However, the elements of the matrix A are generally complex.

It may be noted that the set of all nonsingular homogeneous linear transformations in an n-dimensional complex space is also described by the same matrix equation

$$x' = Ax, \qquad |A| \neq 0$$

However, A is now an n × n matrix, whereas x and x′ are column vectors each having n elements. Proceeding along the same lines as before, it can be shown that this set of all nonsingular linear transformations in an n-dimensional complex space also forms a group known as the *general linear group in a complex space of n dimensions* and is denoted by GL(n, C) or GL(n).

PROBLEM 3.2
The set of all homogeneous nonsingular linear transformations in a complex space of n dimensions forms a group under successive application of transformations. Show that the set of the corresponding transformation matrices also forms a group but under matrix multiplication and is isomorphic to the group of homogeneous linear transformations.

Since, by virtue of Problem 3.2, the group of homogeneous nonsingular linear transformations

$$T_1: x' = Ax, \qquad |A| \neq 0$$
$$T_2: x'' = Bx', \qquad |B| \neq 0$$
$$\cdots\cdots\cdots\cdots\cdots\cdots$$

is isomorphic to the group of the corresponding nonsingular transformation matrices

$$A, B, C, \ldots.$$

the two groups have the same structure. Therefore, while considering the group of nonsingular linear transformations in a space of n dimensions, for convenience in calculations we will make use of the group theoretical properties of the corresponding nonsingular transformation matrices. The group theoretical results thus obtained will be valid for the corresponding group of linear transformations.

Let us next consider that subset of GL(n), which consists of only transformations for which every transformation matrix A is unitary, that is, $A^\dagger A = I = AA^\dagger$, where $A^\dagger$ is the Hermitian conjugate of A and I is the $n \times n$ unit matrix. We will show that this subset forms a subgroup of GL(n). To prove this, we make use Theorem 1.1. According to this theorem: If for the elements P, Q belonging to a subset of a group, the product PQ^{-1}, *under the same law of composition as for the group*, also belongs to the subset, then the subset is a subgroup of the group.

Consider the subset of unitary transformation matrices P, Q, ... so that $P^\dagger P = I$, $Q^\dagger Q = I$, etc. Let us find out whether the product PQ^{-1} also belongs to this subset. If this is the case, then the subset will be a subgroup of the given group. We have

$$(PQ^{-1})^\dagger(PQ^{-1}) = (Q^{-1})^\dagger P^\dagger PQ^{-1} = (Q^\dagger)^{-1} IQ^{-1}, \text{ because P is unitary}$$
$$= (Q^\dagger)^{-1} Q^{-1} = (QQ^\dagger)^{-1} = I^{-1} = I$$

Thus, PQ^{-1} is unitary and hence belongs to the subset. This proves that the subset is a subgroup of GL(n). Of course, this subgroup itself forms a group under the same law of composition as for the group. Since this group is isomorphic to the group of unitary transformations, these transformations also form a group but under successive application of transformations in a complex space of n dimensions. This group of unitary transformations is called *unitary group* in a complex space of n dimensions and is denoted by U(n).

Let us next consider another subset of the group GL(n) that consists only of the transformations for which the transformation matrices are unimodular: $|A| = 1$. Let R, S, ... be the members of this subset so that $|R| = 1$, $|S| = 1$, ... Then $|RS^{-1}| = |R|\ |S^{-1}| = |S^{-1}|$. As $SS^{-1} = I$, we have $|S|\ |S^{-1}| = |I| = 1$. This yields $|S^{-1}| = 1/|S| = 1$. Hence, $|RS^{-1}| = 1$; that is, the product RS^{-1} also belongs to the subset. This shows that the subset R, S, ..., with $|R| = 1$, $|S| = 1$, ... is a subgroup of GL(n). This group of unimodular transformations is known as *special linear group* in a complex space of n dimensions and is denoted by SL(n).

It can be shown in the same manner that the subset consisting only of transformations of SL(n) for which every transformation matrix is unitary is itself a group of linear transformations. This group is known as *special unitary group* or *unimodular unitary group* in a complex space of n dimensions and is denoted by SU(n). The same group is again obtained by applying the condition det A = 1, where A is the transformation matrix, in the choice of a subset of the unitary group U(n).

Using $\supset$ as the symbol for "contains as a subset," we can write

$$GL(n) \supset U(n) \supset SU(n)$$

and

$$GL(n) \supset SL(n) \supset SU(n)$$

We can present these results for the groups of transformations in an n-dimensional space as shown in Figure 3.2.

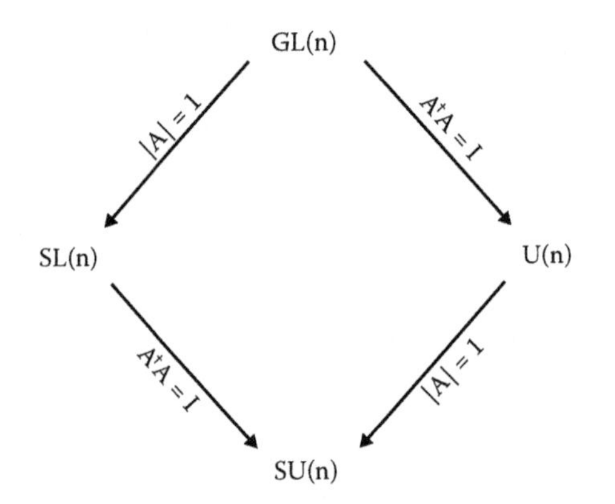

FIGURE 3.2
GL(n) and its subsets as groups.

PROBLEM 3.3
Show that, under successive applications of transformations, the standard Lorentz transformations form a group.

PROBLEM 3.4
Show that the set of transformation matrices of the standard Lorentz transformations forms a group under matrix multiplication and that this group is isomorphic to the group of standard Lorentz transformations.

PROBLEM 3.5
Show that the group of standard Lorentz transformations is bounded but not closed. Is it a compact group?

Since all the parameters constituting the nonsingular transformation matrix A are allowed to vary continuously, the groups of homogeneous nonsingular linear transformations

$$x' = Ax, \qquad |A| \neq 0$$

with various constraints on A are continuous groups.

PROBLEM 3.6
Show that the set of all nonsingular inhomogeneous linear transformations in a complex space of n dimensions forms a group under successive application of transformations.

PROBLEM 3.7
Show that the infinite set of matrices

$$\begin{bmatrix} \cos\varphi & -\sin\varphi \\ \sin\varphi & \cos\varphi \end{bmatrix}$$

where φ varies continuously between zero and 2π, forms a continuous group.

3.3 Order of a Group of Transformations

Let us next find the orders of various groups of transformations. First we consider GL(n), the general linear group in a complex space of n dimensions:

$$x' = Ax, \qquad |A| \neq 0$$

Each transformation matrix A in an n-dimensional space is of order n and therefore has n^2 elements. Since the space is complex, each element involves two continuously varying real parameters: the real and imaginary parts of a complex parameter. Consequently, the total number of real parameters is $2n^2$. Since no relationship is imposed on these parameters, the number of essential parameters is also $2n^2$. This is the order of the group GL(n).

Next we consider the unitary group U(n) in a complex space of n dimensions:

$$x' = Ax, \qquad A^{\dagger}A = I$$

where $A^{\dagger}$ is the Hermitian conjugate of the square matrix A, and I is the $n \times n$ unit matrix. The transformation matrix A will have $2n^2$ continuously varying real parameters. These are not all independent but are related by the matrix equation

$$A^{\dagger}A = I \tag{3.4}$$

Considering the ij-element on both sides of this matrix equation, we get

$$(A^{\dagger}A)_{ij} = (I)_{ij}$$

or

$$\sum_k A^{\dagger}_{ik}A_{kj} = \delta_{ij}$$

where δ_{ij} is the Kronecker delta and is by definition either equal to 1 or zero according as $j = i$ or $j \neq i$. Since $A^{\dagger} \equiv (A^*)^T$, performing the transpose operation, we get

$$\sum_k A^*_{ki}A_{kj} = \delta_{ij}$$

For *diagonal elements*, $j = i$ and the previous equation yields

$$\sum_k A^*_{ki}A_{ki} = 1$$

or

$$\sum_k |A_{ki}|^2 = 1$$

This equation consists of real numbers only. Therefore, for each diagonal element of the product $A^{\dagger}A$, only one equation containing real parameters is involved. Since there are n diagonal elements, the total number of such relations is also n.

For nondiagonal elements, $j \neq i$, and Equation (3.4) yields

$$\sum_k A^*_{ki} A_{kj} = 0 \tag{3.5}$$

This relation involves complex numbers and therefore for various values of j and i, with $j \neq i$, each relation gives two equations in terms of real parameters. Since the total number of nondiagonal elements is $n^2 - n$, the total number of equations relating the real parameters is $2(n^2 - n)$. But, as shown next, these relations are not all independent. This is because for the ij-element of the product $A^\dagger A$, with $j \neq i$, Relation (3.5) is

$$\sum_k A^*_{ki} \; A_{kj} = 0 \tag{3.5'}$$

For the ji-element, with $j \neq i$, Relation (3.5) gives

$$\sum_k A_{kj} A^*_{ki} = 0$$

Taking the complex conjugate of both sides of this equation, we get

$$\sum_k A_{kj} A^*_{ki} = 0$$

or

$$\sum_k A^*_{ki} \; A_{kj} = 0 \tag{3.6}$$

Equation (3.6) is the same as Equation (3.5′). That is, Equation (3.6) for the ji-element does not give any relationship between the parameters different from that given by Equation (3.5′) for the ij-element. Thus, the total number of independent relationships between the real parameters of nondiagonal elements is $(n^2 - n)$. Consequently, the total number of independent relations between real parameters of the matrix A is $\{n + (n^2 - n)\} \equiv n^2$. Hence, the maximum number of *independent* real parameters, that is, the number of essential parameters of the group U(n), is $(2n^2 - n^2) \equiv n^2$.

PROBLEM 3.8

Show that the orders of the groups of transformations SU(n) and SL(n) are $n^2 - 1$ and $2n^2 - 1$, respectively.

3.4 Lie Groups

For the sake of simplicity, let us first consider a continuous group G characterized with only one continuously varying real parameter, say α, in a one-dimensional linear vector space. Then the group elements corresponding to the real values a, b, c, ... of the parameter α may be denoted by g(a), g(b), g(c), ... Since G is a group, it must possess the closure property; the product of its elements should also be an element of the group. Therefore, if g(a) and g(b) are two group elements corresponding to arbitrary values a and b of the parameter α, then there must exist a parameter value c such that

$$g(a)\, g(b) = g(c)$$

Since this must hold for all elements of G, we should have a set of parameter values c for all possible sets of parameter values a and b. Thus, there must exist a functional relation among a, b, and c, and we may write

$$c = \varphi(a, b)$$

Since the real values of a and b vary continuously, the parameter c is a real function of real parameters a and b. As a group always contains the identity element, there must exist a parameter value e such that g(e) is the identity of G. Moreover, as each group element must have an inverse, with each parameter value a, there is associated a parameter value $\bar{a}$ such that

$$g(a)g(\bar{a}) = g(e) = g(\bar{a})g(a)$$

If c is an analytic function of a and b—that is, if $\varphi(a, b)$ can be differentiated any number of times with respect to its arguments and $\bar{a}$ is an analytic function of a—then the continuous group G is called a *Lie group*. Sophus Lie was a Norwegian mathematician, and to honor his name such groups are called Lie groups. The definition can be easily extended to a continuous group of r essential parameters in a vector space of n dimensions.

We will now consider a very simple example of a Lie group. Consider a one-parameter continuous group of homogeneous linear transformations in a one-dimensional real vector space, the transformation equations being of the type

$$x' = \alpha x, \qquad \alpha \neq 0$$

with the real parameter α varying continuously. The law of composition is successive application of transformations. Let us consider any two members of the group of transformations characterized by the parameter values, say, $\alpha = a$ and b:

$$x' = ax, \qquad a \neq 0$$

$$x'' = bx', \qquad b \neq 0$$

The successive application of these transformations must yield another member of this group:

$$x'' = bx' = bax = cx, \quad c \neq 0$$

where $c = ba$.

The inverse of the transformation

$$x' = ax, \quad a \neq 0$$

must be

$$x = \bar{a}x' = \frac{1}{a}x', \qquad \bar{a} \neq 0$$

where

$$\bar{a} = \frac{1}{a}$$

Since c (= ba) is an analytical function of a and b—that is, since we can differentiate ba with respect to a and b any number of times—and similarly $\bar{a}$ is an analytic function of a, this continuous group is a Lie group.

PROBLEM 3.9
Show that the continuous group of inhomogeneous linear transformations

$$x' = x + a$$

where a is a continuously varying real parameter, is a Lie group under successive application of transformations.

PROBLEM 3.10
Show that the two-parameter group of inhomogeneous linear transformations

$$x' = ax + b, \quad a \neq 0$$

is a non-Abelian Lie group under successive application of transformations.

Let us now generalize. Consider an r-parameter continuous group of linear transformations in a real vector space of n dimensions, with $x_1, x_2, \ldots, x_n$ as basis vectors and with $a_1, a_2, \ldots, a_r$ as continuously varying real parameters. Then we may write the transformations as

$$\begin{aligned} x'_1 &= f_1(x_1, x_2, \ldots, x_n; a_1, a_2, \ldots, a_r) \\ x'_2 &= f_2(x_1, x_2, \ldots, x_n; a_1, a_2, \ldots, a_r) \\ &\cdots\cdots\cdots\cdots\cdots\cdots \\ x'_n &= f_n(x_1, x_2, \ldots, x_n; a_1, a_2, \ldots, a_r) \end{aligned} \tag{3.7}$$

Symbolically, we may write Equations (3.7) as

$$x' = f(x, a) \tag{3.8}$$

where x and a stand for the sets of n variables and r parameters, respectively. The properties of the group of transformations may be expressed as follows:

(i) There exists a set of parameter values e such that

$$x' = f(x, e) = x$$

This is the identity transformation.

(ii) For every set of parameter values a, there exists a unique set of parameter values $\bar{a}$ such that for $x' = f(x, a)$, there exists $x = f'(x', \bar{a})$ that when multiplied to the given transformation yields the identity transformation. Then each one of the transformations is the inverse of the other. Since Equation (3.8) must be solvable for x in terms of x′, the Jacobian must be different from zero:

$$\begin{vmatrix} \frac{\partial f_1}{\partial x_1} & \frac{\partial f_1}{\partial x_2} & \cdots & \frac{\partial f_1}{\partial x_n} \\ \frac{\partial f_2}{\partial x_1} & \frac{\partial f_2}{\partial x_2} & \cdots & \frac{\partial f_2}{\partial x_n} \\ \vdots & \vdots & \cdots & \vdots \\ \frac{\partial f_n}{\partial x_1} & \frac{\partial f_n}{\partial x_2} & \cdots & \frac{\partial f_n}{\partial x_n} \end{vmatrix} \neq 0$$

(iii) The product of two members of the group is a member of the group. Consider two transformations of the group:

$$x' = f(x, a)$$

$$x'' = f(x', b)$$

If we perform them in succession, we get

$$x'' = f[f(x, a), b] = f(x, c)$$

This relation determines c in terms of a and b:

$$c = \varphi(a, b)$$

(iv) Of course, for linear transformations, the multiplication is always associative.

If c is an analytic function of a and b, and $\bar{a}$ is an analytic function of a, then, by definition, the previously given continuous group is a Lie group of transformations.

3.5 Generators of Lie Groups

Sophus Lie has shown that all elements of a Lie group that can be reached continuously from its identity can be obtained by using a particular set of operators. These operators are called *generators of the Lie group*. In this section, we will find mathematical expressions for these generators and examine their role in the study of the local properties of a Lie group, that is, the properties in the neighborhood of its identity.

Consider first a one-parameter Lie group of transformations

$$x' = f(x, a) \tag{3.9}$$

in a one-dimensional coordinate space, the law of composition being successive application of transformations.

The groups of transformations

(1) $x' = ax, \quad a \neq 0$

and

(2) $x' = x + a$

are two examples of such transformations. The first is a homogeneous, whereas the second is an inhomogeneous group of linear transformations.

Let $a = e$ be the value of the parameter that gives the identity transformation $x' = x$. Then

$$x' = f(x, e) = x \tag{3.10}$$

In examples (1) and (2), the identity transformation $x' = x$ is obtained by choosing the parameter values as $a = 1$ and $a = 0$, respectively.

The *infinitesimal transformation* is defined as what differs infinitesimally from the identity transformation. Therefore, the infinitesimal transformation is obtained by making a small change $da \equiv \varepsilon$ in the parameter value e for the identity transformation. This will cause a small change in x, and Equation (3.10) will yield

$$x' = x + dx = f(x, e + da) \tag{3.11}$$

Expanding the right-hand side of Equation (3.11) to the lowest order in da, we get

$$x + dx = f(x,e) + \left[\frac{df(x,a)}{da}\right]_{a=e} da = f(x,e) + \frac{df(x,e)}{da} da \tag{3.12}$$

where we have written

$$\frac{df(x,e)}{da} \equiv \left[\frac{df(x,a)}{da}\right]_{a=e}$$

Substituting the expression for f(x,e) from Equation (3.10) in Equation (3.12), we get

$$dx = \frac{df(x,e)}{da} da$$

$$dx = u(x)da \tag{3.13a}$$

where u(x) stands for df(x, e)/da. By virtue of Equation (3.11) and writing $\in$ for da, Equation (3.13a) takes the form

$$x' - x = u(x) \in \tag{3.13b}$$

Each of Equations (3.13a) and (3.13b) represents the infinitesimal transformation.

Now let F be an arbitrary function of x:

$$F = F(x)$$

Then a change in F under an infinitesimal change dx in x is given by

$$dF = \frac{dF}{dx} dx \tag{3.14}$$

Eliminating dx between Equations (3.13a) and (3.14), we get

$$dF = \frac{dF}{dx} u(x)da$$

$$dF = da\left[u(x)\frac{d}{dx}\right]F \tag{3.15}$$

Denoting the operator $\left[u(x)\frac{d}{dx}\right]$ by L, we have

$$L = u(x)\frac{d}{dx} \tag{3.16}$$

Then Equation (3.15) may be written as

$$dF = daLF$$

This equation yields

$$d \equiv daL \tag{3.17}$$

Now if g(e) is the identity element of a continuous group G, then the element g(e + da), close to g(e), is obtained by making a small change da in the parameter value a = e and is given by

$$\begin{aligned} g(e + da) &= g(e) + dg(e) \\ &= g(e) + daLg(e) \\ &= (1 + daL)g(e) \end{aligned} \tag{3.18}$$

where the result obtained in Equation (3.17) has been used. Equation (3.18) shows that by an application of the operator (1 + daL) on the identity element g(e), the element g(e + da) close to the identity is generated. By successive applications of the operator (1 + daL), we can obtain an element g(e + a) that can be reached continuously from the identity. This is achieved in the following manner. Applying this operator N times, we have

$$g(e + Nda) = (1 + daL)^N g(e) = (1 + NdaL/N)^N g(e)$$

Taking the limit as N goes to infinity, we get

$$g(e + a) = \lim_{N \to \infty} (1 + aL/N)Ng(e) \tag{3.19a}$$

as lim N da = a. Equation 3.19a now becomes

$N \to \infty$

$$g(e + a) = [\exp (aL)]g(e) \tag{3.19b}$$

Equation 3.19b shows that any element of a Lie group G that can be reached continuously from its identity can be obtained from the identity by using the operator

$$L \equiv u(x)\frac{d}{dx}$$

The operator L is therefore called the *generator of the group*. All the elements of G that are connected to its identity can be obtained using this generator.

Example 3.1

Consider a one-parameter group of homogeneous linear transformations in a one-dimensional real vector space. This group is described by

$$x' = ax, \quad a \neq 0$$

where the real parameter a is allowed to vary continuously. The identity transformation, $x' = x$, is then obtained by putting $a = 1$. The infinitesimal transformation is obtained by writing $a = 1 + \varepsilon$ and is consequently given by

$$x' = (1 + \varepsilon)x$$

or

$$x' - x = \varepsilon x \tag{3.20}$$

By virtue of Equation (3.13b), we have

$$x' - x = \varepsilon\, u(x) \tag{3.21}$$

Comparing Equations (3.20) and (3.21), we get

$$u(x) = x$$

Hence, by virtue of Equation (3.16), the only generator L of this group is given by

$$L = u(x)\frac{d}{dx} = x\frac{d}{dx}$$

PROBLEM 3.11
Find the generator of the group of inhomogeneous linear transformations in a one-dimensional real vector space:

$$x' = x + a$$

The previous discussion can easily be extended to an r-parameter Lie group of transformations in n variables $x_1, x_2, \ldots, x_n$. The transformation equations are

$$x'_1 = f_1(x_1, x_2, \ldots, x_n; a_1, a_2, \ldots, a_r)$$
$$x'_2 = f_2(x_1, x_2, \ldots, x_n; a_1, a_2, \ldots, a_r)$$
$$\cdots\cdots\cdots\cdots\cdots\cdots\cdots\cdots$$
$$x'_n = f_n(x_1, x_2, \ldots, x_n; a_1, a_2, \ldots, a_r)$$

where $a_1, a_2, \ldots, a_r$ are essential parameters. This set of transformation equations is symbolically written as

$$x'_i = f_i(x, a), \quad i = 1, 2, \ldots, n \tag{3.22}$$

where x stands for the set of variables $x_1, x_2, \ldots, x_n$, and a stands for the set of essential parameters $a_1, a_2, \ldots, a_r$. The identity transformation is given by

$$x'_i = f_i(x, e) = x_i$$

A small change in x_i can be produced by a small change in e:

$$x'_i = x_i + dx_i = f_i(x; e + da)$$

that is,

$$x'_i = x_i + dx_i = f_i(x_1, x_2, \ldots, x_n; e_1 + da_1, e_2 + da_2, \ldots, e_r + da_r)$$

Expanding the right-hand side of this equation to the lowest order in da_k, we get

$$dx_i = \sum_{k=1}^{r} \frac{\partial f_i(x,e)}{\partial a_k} da_k$$

$$= \sum_{k=1}^{r} u_{ik}(x) da_k, \quad i = 1, 2, \ldots, n \tag{3.23}$$

where

$$u_{ik}(x) = \frac{\partial f_i(x,e)}{\partial a_k}$$

If we use the summation convention, we can drop the sigma sign and write Equation (3.23) as

$$dx_i = u_{ik}(x)\, da_k, \qquad i = 1, 2, \ldots, n \tag{3.24}$$

where the sum over k runs from 1 to r. We can write this equation also as

$$x'_i - x_i = u_{ik}(x)\varepsilon_k, \qquad i = 1, 2, \ldots, n \tag{3.25}$$

and the sum over k is from 1 to r. This set of equations represents the infinitesimal transformation.

Let us now consider an arbitrary function F of n variables $x_1, x_2, \ldots, x_n$:

$$F = F(x).$$

Then

$$dF = \sum_{i=1}^{n} \frac{\partial F(x)}{\partial x_i} dx_i \tag{3.26}$$

Substituting the expression for dx_i from Equation (3.24) in Equation (3.26), we obtain

$$dF = \sum_{i=1}^{n} \frac{\partial F(x)}{\partial x_i} \sum_{k=1}^{r} u_{ik}(x) da_k$$

or

$$dF = \sum_{k=1}^{r} da_k \sum_{i=1}^{n} \left[u_{ik}(x) \frac{\partial}{\partial x_i} \right] F(x)$$

$$= \sum_{k=1}^{r} da_k L_k F(x)$$

where

$$L_k = \sum_{i=1}^{n} u_{ik}(x) \frac{\partial}{\partial x_i}, \qquad k = 1, 2, \ldots, r \tag{3.27}$$

are the generators of an r-parameter group. It may be emphasized that the number of generators of a group is the same as the number of its essential parameters. It can be proved that these generators are linearly independent.

Using the summation convention, Equation (3.27) can be written as

$$L_k = u_{ik}(x) \frac{\partial}{\partial x_1} \quad k = 1, 2, \ldots, r \tag{3.28}$$

The summation is over i that runs from 1 to n.

Thus, if we know the equations for the infinitesimal transformation of a group, we can read the values of $u_{ik}(x)$ by comparison with Equation (3.25) and then calculate the generators of the group by using Formula (3.28).

Let us now calculate the generators of various Lie groups of transformations. In further analysis, unless otherwise stated, a group or a continuous group will mean a Lie group.

Example 3.2

Let us find the generators of the two-parameter group of inhomogeneous linear transformations in a real vector space of one dimension:

$$x' = a\,x + b, \quad a \neq 0 \tag{3.29}$$

The identity transformation, $x' = x$, corresponds to $a = 1$, $b = 0$. Therefore, the infinitesimal transformation is obtained by the substitution: $a = 1 + \varepsilon_1$, $b = \varepsilon_2$. This yields

$$x' = (1 + \varepsilon_1)x + \varepsilon_2$$

Writing x_1 and x_1' for x and x', respectively, and simplifying, we get

$$x'_1 - x_1 = \varepsilon_1 x_1 + \varepsilon_2 \tag{3.30}$$

But from Equation (3.25), we have

$$x'_i - x_i = \varepsilon_k u_{ik}, \qquad i = 1, 2, \ldots, n \tag{3.31a}$$

The summation is over k that runs from 1 to r. For a one-dimensional space, $n = 1$ and the previous equation yields

$$x'_1 - x_1 = \varepsilon_k u_{1k} \tag{3.31b}$$

For a two-parameter group, $r = 2$, and consequently Equation (3.31b) yields

$$x'_1 - x_1 = \varepsilon_1 u_{11} + \varepsilon_2 u_{12} \tag{3.31c}$$

Comparing Equations (3.30) and (3.31c), we get

$$u_{11} = x_1, \; u_{12} = 1$$

By virtue of Equation (3.28), the generators L_k of the group are given by

$$L_k = u_{ik} \frac{\partial}{\partial x_i}, \qquad k = 1, 2 \tag{3.31d}$$

For a group of transformations in a one-dimensional vector space, as $n = 1$, i can take only one value. Therefore, Equation 3.31d gives

$$L_k = u_{ik} \frac{d}{dx_1}, \qquad k = 1, 2$$

For k = 1:
$$L_1 = u_{11} \frac{d}{dx_1} = x_1 \frac{d}{dx_1} \equiv x \frac{d}{dx}$$

For k = 2:
$$L_2 = u_{12} \frac{d}{dx_1} = \frac{d}{dx_1} \equiv \frac{d}{dx}$$

These are the two generators of this group.

Let us find the commutator of these two generators. Let the commutator operate upon an arbitrary function φ. Then

$$[L_1, L_2]\varphi = \left[x \frac{d}{dx}, \frac{d}{dx}\right]\varphi = x \frac{d}{dx}\left(\frac{d\varphi}{dx}\right) - \frac{d}{dx}\left(x \frac{d\varphi}{dx}\right)$$

$$= x \frac{d^2\varphi}{dx^2} - \frac{d\varphi}{dx} - x \frac{d^2\varphi}{dx^2} = \frac{d}{dx}\varphi = -L_2\varphi$$

This yields

$$[L_1, L_2] = -L_2$$

Example 3.3

We find the generators of general linear group in a real vector space of two dimensions. This group is denoted by GL(2,R). The transformation equations are

$$x_1' = a_{11}x_1 + a_{12}x_2 \qquad \begin{vmatrix} a_{11} & a_{12} \\ a_{21} & a_{22} \end{vmatrix} \neq 0$$
$$x_2' = a_{21}x_1 + a_{22}x_2$$

The identity transformation is obtained by choosing $a_{11} = 1$, $a_{12} = 0$, $a_{21} = 0$, $a_{22} = 1$. Therefore, the infinitesimal transformation is obtained by writing $a_{11} = 1 + \varepsilon_1$, $a_{12} = \varepsilon_2$, $a_{21} = \varepsilon_3$, $a_{22} = 1 + \varepsilon_4$. This yields

$$x_1' = (1 + \varepsilon_1)\, x_1 + \varepsilon_2 x_2$$
$$x_2' = \varepsilon_3 x_1 + (1 + \varepsilon_4) x_2$$

or

$$x_1' - x_1 = \varepsilon_1 x_1 + \varepsilon_2\, x_2$$
$$x_2' - x_2 = \varepsilon_3 x_1 + \varepsilon_4 x_2 \tag{3.32a}$$

Now, from Equation (3.25), we have

$$x_i' - x_i = \varepsilon_k u_{ik}, \quad i = 1, 2, \ldots, n$$
$$k = 1, 2, \ldots, r$$

For $r = 4$, this equation gives

$$x_i' - x_i = \varepsilon_1 u_{i1} + \varepsilon_2 u_{i2} + \varepsilon_3 u_{i3} + \varepsilon_4 u_{i4} \tag{3.32b}$$

For $i = 1$ and $i = 2$, Equation (3.32b) yields

$$x_1' - x_1 = \varepsilon_1 u_{11} + \varepsilon_2 u_{12} + \varepsilon_3 u_{13} + \varepsilon_4 u_{14} \tag{3.33a}$$
$$x_2' - x_2 = \varepsilon_1 u_{21} + \varepsilon_2 u_{22} + \varepsilon_3 u_{23} + \varepsilon_4 u_{24} \tag{3.33b}$$

Comparing Equations (3.32) and (3.33), we get

$$u_{11} = x_1,\ u_{12} = x_2,\ u_{13} = u_{14} = 0$$
$$u_{21} = u_{22} = 0,\ u_{23} = x_1,\ u_{24} = x_2$$

For n = 2, as the index i runs from 1 to 2, the Formula (3.28) for the generators of a Lie group takes the form

$$L_k = u_{ik}\frac{\partial}{\partial x_i} = u_{1k}\frac{\partial}{\partial x_1} + u_{2k}\frac{\partial}{\partial x_2}, \qquad k = 1, 2, \ldots, r$$

Therefore, the four generators are given by

$$L_1 = u_{11}\frac{\partial}{\partial x_1} + u_{21}\frac{\partial}{\partial x_2} = x_1\frac{\partial}{\partial x_1} \equiv x\frac{\partial}{\partial x}$$

$$L_2 = u_{12}\frac{\partial}{\partial x_1} + u_{22}\frac{\partial}{\partial x_2} = x_2\frac{\partial}{\partial x_1} \equiv y\frac{\partial}{\partial x}$$

$$L_3 = u_{13}\frac{\partial}{\partial x_1} + u_{23}\frac{\partial}{\partial x_2} = x_1\frac{\partial}{\partial x_2} \equiv x\frac{\partial}{\partial y}$$

$$L_4 = u_{14}\frac{\partial}{\partial x_1} + u_{24}\frac{\partial}{\partial x_2} = x_2\frac{\partial}{\partial x_2} \equiv y\frac{\partial}{\partial y}$$

The same result can be obtained by writing the transformation equations for the group GL(2,R) in matrix form:

$$x' = Ax, \qquad |A| \neq 0$$

The transformation matrix A is a nonsingular 2 × 2 matrix. The identity transformation is obtained by taking A = I:

$$x' = Ix, \qquad |I| = 1 \neq 0$$

The infinitesimal transformation is obtained by writing A = I + B, where the matrix B is a 2 × 2 matrix and has all its elements in the vicinity of zero:

$$x' = (I + B)x, \qquad |I + B| \neq 0$$

The matrix B may therefore be written as

$$B = \begin{bmatrix} \varepsilon_1 & \varepsilon_2 \\ \varepsilon_3 & \varepsilon_4 \end{bmatrix}$$

Since we are considering the transformations in a two-dimensional real vector space, the transformation matrices must be 2 × 2, and the column vectors should have two elements each. Therefore, the previous matrix equation can be written as

$$\begin{bmatrix} x_1' \\ x_2' \end{bmatrix} = \begin{bmatrix} 1+\varepsilon_1 & \varepsilon_2 \\ \varepsilon_3 & 1+\varepsilon_4 \end{bmatrix} \begin{bmatrix} x_1 \\ x_2 \end{bmatrix}$$

Multiplying the square matrix and the column vector on the right side of this matrix equation and then comparing the corresponding elements on the two sides of the equation, we get

$$x_1\ (1+\varepsilon_1)x_1 + \varepsilon_2 x_2$$

$$x_2\ \varepsilon_3 x_1 + (1+\varepsilon_4)x_2$$

The rest of the procedure is the same as given already.

PROBLEM 3.12
Prove that for the group of transformations given in Example 3.3:

$$[L_1, L_2] = -L_2; \quad [L_1, L_3] = L_3; \quad [L_1, L_4] = 0$$

and

$$[L_2, L_3] = L_4 - L_1; \quad [L_2, L_4] = -L_2; \quad [L_3, L_4] = L_3$$

(Notice that the commutator of any two generators of a group is a linear combination of the generators of the group.)

PROBLEM 3.13
Find the generators of the following two-parameter group of homogeneous linear transformations in a real space of two dimensions:

$$x' = ax, \quad a \neq 0$$

$$y' = by, \quad b \neq 0$$

3.6 Real Orthogonal Group in Two Dimensions: O(2)

The definition of an orthogonal matrix varies in literature. It is therefore necessary to first define it and then proceed further. In literature on pure mathematics, an orthogonal matrix A is usually defined as a square matrix in the field of real numbers such that

$$A^T A = I$$

where A^T is the transpose of the matrix A. In that case, the orthogonal transformations defined by

$$x' = Ax, \quad A^T A = I$$

form a group over the field of real numbers in a space of n dimensions. This group is denoted by O(n). Notice that by virtue of this definition of an orthogonal matrix, the group O(n) is always over a field of real numbers.

On the other hand, in physics, a square matrix A, whether its elements are real or complex, is said to be orthogonal if it satisfies the equation

$$A^T A = I$$

An orthogonal group is now defined as a group of linear transformations

$$x' = Ax$$

in a vector space (which may be real or complex) of n dimensions such that the transformation matrix A is orthogonal:

$$A^T A = I$$

If the vector space is real (i.e., the elements of A are all real), then in the usual notation we should label the group as O(n,R); it is a real orthogonal group. However, it is conventional in this particular case to deviate from this notation and write it as O(n) instead of O(n,R), because, in pure mathematics, the orthogonal matrix is defined only over a real field. Hence, O(n) stands for a group of orthogonal transformations in a real vector space of n dimensions.

Let us now consider the set of equations underlying the group of general linear homogeneous transformations in a real vector space of two dimensions, viz.

$$\begin{aligned} x_1' &= a_{11}x_1 + a_{12}x_2 \\ x_2' &= a_{21}x_1 + a_{22}x_2 \end{aligned} \qquad \begin{vmatrix} a_{11} & a_{12} \\ a_{21} & a_{22} \end{vmatrix} \neq 0 \tag{3.34}$$

Out of these, we consider only transformations that leave

$$x^2_1 + x^2_2$$

invariant:

$$x_1'^2 + x_2'^2 = x_1^2 + x_2^2 \tag{3.35}$$

Substituting the expressions for x_1' and x_2' from Equations (3.34) in Equation (3.35), we get

$$(a_{11}x_1 + a_{12}x_2)^2 + (a_{21}x_1 + a_{22}x_2)^2 = x^2_1 + x^2_2$$

or

$$a^2_{11}\, x^2_1 + a^2_{12}x^2_2 + 2\, a_{11}a_{12}x_1x_2 + a^2_{21}x^2_1 + a^2_{22}x^2_2 + 2a_{21}a_{22}\, x_1x_2 = x^2_1 + x^2_2$$

Comparing the coefficients of x^2_1, x^2_2, and x_1x_2 on the two sides of this equation, we get

$$\begin{aligned} a^2_{11} + a^2_{21} &= 1 \\ a^2_{12} + a^2_{22} &= 1 \\ a_{11}a_{12} + a_{21}a_{22} &= 0 \end{aligned} \tag{3.36}$$

Thus, by virtue of Equation (3.35), the four parameters of Equations (3.34) are subjected to the previous three conditions. Hence, the group of transformations

$$\begin{aligned} x_1' &= a_{11}x_1 + a_{12}x_2 \\ x_2' &= a_{21}x_1 + a_{22}x_2 \end{aligned} \qquad \begin{vmatrix} a_{11} & a_{12} \\ a_{21} & a_{22} \end{vmatrix} \neq 0$$

which leave $x^2_1 + x^2_2$ invariant is actually a one-parameter group. This group is called the *orthogonal group* in a (real) vector space of two dimensions and is denoted by O(2). All the conditions (3.36) are satisfied if we choose

$$\begin{aligned} a_{11} &= \cos\varphi \\ a_{12} &= -\sin\varphi \\ a_{21} &= \sin\varphi \\ a_{22} &= \cos\varphi \end{aligned}$$

The transformation equations for the group then can be written as

$$\begin{aligned} x_1' &= x_1 \cos\varphi - x_2 \sin\varphi \quad 0 \le \varphi \le 2\pi \\ x_2' &= x_1 \sin\varphi + x_2 \cos\varphi \end{aligned} \tag{3.37}$$

where φ is the angle of rotation about x_3-axis. The corresponding transformation matrix is

$$\begin{bmatrix} \cos\varphi & -\sin\varphi \\ \sin\varphi & \cos\varphi \end{bmatrix}$$

It is called the *rotation matrix*, is denoted by R(φ), and represents a group element in general. The generator of this group O(2) is obtained by proceeding in the usual way.

PROBLEM 3.14

Show that Equations (3.36) could also be satisfied by choosing the transformation matrix as

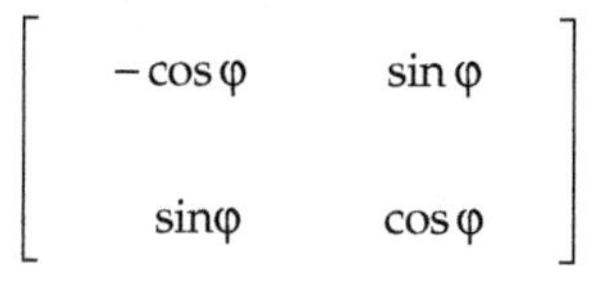

What is its determinant equal to?

We will now determine the generator of this very group by using matrix notation. The general linear group GL(n,R) in a real vector space consists of linear transformations of the form

$$x' = Ax, \qquad |A| \neq 0 \tag{3.38}$$

In a two-dimensional real space, out of these, we have to choose those transformations which leave $x^2_1 + x^2_2$ invariant. In matrix notation,

$$x_1^2 + x_2^2 = [x_1 \quad x_2]\begin{bmatrix} x_1 \\ x_2 \end{bmatrix} = x^T x$$

Thus, we have to choose transformations for which

$$x'^T x' = x^T I x \tag{3.39}$$

Substituting the expression for x′ from Equation (3.38) in Equation (3.39), we get

$$(Ax)^T Ax = x^T I x$$

or

$$x^T A^T A x = x^T I x$$

So that the equation may be valid for all x, we must have

$$A^T A = I \tag{3.40}$$

That is, the matrix A must be orthogonal. It is for this reason that the group is called an orthogonal group. Hence, the group O(2) is defined by

$$x' = Ax, \quad A^T A = I \tag{3.41}$$

where the elements of 2×2 matrix A are real. The identity transformation is obtained by putting $A = I$, and hence the infinitesimal transformation is obtained by writing $A = I + B$, where the 2×2 matrix B has all its elements in the neighborhood of zero:

$$x' = (I + B)\, x \tag{3.42}$$

As the transformation matrix $I + B$ for the infinitesimal transformation must be orthogonal, we have

$$(I + B)^T (I + B) = I$$

or

$$(I + B^T)(I + B) = I$$

or

$$I + B + B^T + B^T B = I$$

or

$$B^T = -B$$

$B^T B$ is neglected compared with B and B^T as its elements will be very small as compared with those of B or B^T. The last equation shows that 2×2 matrix B must be antisymmetric. We may therefore write it as

$$B = \begin{bmatrix} 0 & -\varepsilon_1 \\ \varepsilon_1 & 0 \end{bmatrix} \tag{3.43a}$$

This yields

$$I + B = \begin{bmatrix} 1 & -\varepsilon_1 \\ \varepsilon_1 & 1 \end{bmatrix} \tag{3.43b}$$

indicating that O(2) is a one-parameter group. Substituting the expressions for the vectors x and x′, and for the square matrix B in Equation (3.42) we obtain

$$\begin{bmatrix} x_1' \\ x_2' \end{bmatrix} = \begin{bmatrix} 1 & -\varepsilon \\ \varepsilon_1 & 1 \end{bmatrix} \begin{bmatrix} x_1 \\ x_2 \end{bmatrix}$$

Comparing the corresponding elements on the two sides of this equation, we get

$$x_1' - x_1 = -\varepsilon_1 x_2$$

$$x_2' - x_2 = \varepsilon_1 x_1$$

Proceeding further in accordance with the technique developed earlier, we obtain the generator of O(2), namely,

$$L = x\frac{\partial}{\partial y} - y\frac{\partial}{\partial x}$$

It may be noted that the matrix B could be chosen as

$$B = \begin{bmatrix} 0 & \varepsilon_1 \\ -\varepsilon_1 & 0 \end{bmatrix} \tag{3.43c}$$

This will merely indicate that the rotation about x_3-axis is now in the opposite direction.

PROBLEM 3.15
Find the generators of the groups O(3), U(2), and SL(2,R).

The (real) orthogonal group O(n) is defined by

$$x' = Ax, \qquad A^T A = I$$

The constraint $A^T A = I$ yields

$$|A^T A| = |I| = 1$$

or

$$|A^T|\ |A| = 1$$

or

$$|A|\ |A| = 1$$

or

$$|A|^2 = 1$$

or

$$|A| = \pm 1$$

The group O(n) thus decomposes into two pieces that are disconnected, and we cannot go continuously from one piece to the other. The orthogonal transformations with $|A| = +1$ form a subgroup SO(n), called special orthogonal group in a real space of n dimensions, of the group O(n). The other piece with $|A| = -1$ does not form a subgroup: the product of any two members of this piece has $|A| = +1$ and is therefore not a member of this piece.

The group SO(n) or R(n) defined by

$$x' = Ax, \qquad A^TA = I, \qquad |A| = 1$$

is the group of all rotations in a real space of n dimensions. This is because it keeps $x^2_1 + x^2_2 + \cdots + x^2_n$ invariant and also has $|A| = 1$. Of course, it is the rotation of a position vector in a fixed coordinate system that keeps its magnitude unchanged: only the direction changes. Such rotations are called *proper.* Evidently, SO(3) is the rotational symmetry group in our world of three dimensions.

R(n) has the same number of parameters as O(n). It may be pointed out that R(n) in a space of an odd number of dimensions does not include reflections $x' = -x$, because then the determinant of the transformation matrix is −1.

The group of rotations is compact as it is closed and bounded: the angle of rotation about any axis lies in the interval $0 \le \varphi \le 2\pi$. It may be mentioned that the translation group $x' = x + a$, where the parameter a varies continuously between plus and minus infinity, is not compact because a is unbounded.

PROBLEM 3.16
Show that O(n) is an $\frac{n(n-1)}{2}$ parameter group.

We next define the term *connectivity*. Consider an arbitrary element of an r-parameter continuous group G. Starting from this element, if we can reach the identity element by a continuous variation of r-parameters, then the group G is said to be connected or to possess connectivity. This means that any pair of points in the group space can be connected by an arc generated by a continuous variation of the group parameter.

The rotation group SO(n), a subgroup of O(n), is a connected group, whereas the full orthogonal group O(n) is not. This is because for the group O(n) it is not possible to pass continuously from the orthogonal transformation matrices of determinant +1 to those with determinant equal to –1. A careful consideration shows that the elements of SO(n), the connected piece of O(n) (≡ O(n, R)), generate rotation, whereas the elements of the other disjoint piece generate improper rotations, that is, rotation combined with reflection.

The rotation group SO(2) in two-dimensional real space is a one-parameter compact Abelian Lie group. However, the rotation groups SO(n) where $n \geq 3$ are not Abelian.

PROBLEM 3.17

The group of linear transformations

$$x' = Ax, \qquad |A| \neq 0$$

in a space of two dimensions leaves the real quadratic form $x^2_1 - x^2_2$ invariant. Show that the transformation matrix A is given by

$$A = \begin{bmatrix} \cosh\varphi & \sinh\varphi \\ \sinh\varphi & \cosh\varphi \end{bmatrix}$$

This group is denoted by SO(1,1) indicating that it leaves $x^2_1 - x^2_2$ invariant.

The group O(n) is an example of what is called a *mixed continuous group.* It is an infinite group whose elements are characterized by a set of continuous parameters together with a set of discrete labels. For instance, the elements of the real orthogonal group O(n) are characterized by $n(n-1)/2$ real continuous parameters and the sign of the determinant of the transformation matrix of the element.

3.7 Generators of SU(2)

Let us next calculate the generators of SU(2), the special (unimodular) unitary group in a complex space of two dimensions. The group is characterized by the transformations of the type

$$x_1' = a_{11}x_1 + a_{12}x_2$$

$$x_2' = a_{21}x_1 + a_{22}x_2$$

where the a's are complex numbers, and the transformation matrix

$$A = \begin{bmatrix} a_{11} & a_{12} \\ a_{21} & a_{22} \end{bmatrix}$$

is unimodular, $|A| = 1$, and unitary, $A^{\dagger}A = I$. In matrix form, the transformation equations for the group SU(2) can be written as

$$x' = A\,x$$

where $|A| = 1$ and $A^{\dagger}A = I$.

The identity transformation, $x' = x$, is obtained by putting $A = I$. Therefore, the infinitesimal transformation is given by

$$x' = (I + B)\,x \tag{3.44}$$

where all the elements of the 2×2 matrix B are infinitesimal. Since $I + B$, a transformation matrix, must be unitary, we have

$$(I + B)^{\dagger}(I + B) = I$$

This yields

$$B^{\dagger} = -B \tag{3.45}$$

as $B^{\dagger}B$, being of second order of smallness, has been neglected compared with B or $B^{\dagger}$. Equation (3.45) shows that the matrix B must be anti-Hermitian (also called skew-Hermitian). Let us write

$$B = \begin{bmatrix} a & b \\ c & d \end{bmatrix}$$

Then Equation (3.45) gives

$$\begin{bmatrix} a & b \\ c & d \end{bmatrix} = -\begin{bmatrix} a & b \\ c & d \end{bmatrix}^{\dagger} = -\begin{bmatrix} a^{*} & c^{*} \\ b^{*} & d^{*} \end{bmatrix}$$

Comparing the corresponding elements on the two sides of the previous equation, we get

$$a = -a^*, \quad b = -c^*, \quad c = -b^*, \quad d = -d^*$$

These relations show that the diagonal elements a and d of the matrix B are pure imaginary. The other two relations are identical and show that any one of b and c is just the negative of the complex conjugate of the other. In terms of infinitesimal parameters, we may write $a = i\,\varepsilon_1$, $b = \varepsilon_2 + i\,\varepsilon_3$, $c = -\varepsilon_2 + i\,\varepsilon_3$, and $d = i\,\varepsilon_4$. Then the matrix B can be written as

$$B = \begin{bmatrix} i\varepsilon_1 & \varepsilon_2 + i\varepsilon_3 \\ -\varepsilon_2 + i\varepsilon_3 & i\varepsilon_4 \end{bmatrix} \tag{3.46}$$

Therefore, we have

$$I + B = \begin{bmatrix} 1 + i\varepsilon_1 & \varepsilon_2 + i\varepsilon_3 \\ -\varepsilon_2 + i\varepsilon_3 & 1 + i\varepsilon_4 \end{bmatrix} \tag{3.47}$$

The condition that the matrix must be unimodular gives

$$|I + B| = 1$$

or

$$(1 + i\,\varepsilon_1)(1 + i\,\varepsilon_4) - (\varepsilon_2 + i\,\varepsilon_3)(-\varepsilon_2 + i\,\varepsilon_3) = 1$$

which, to the first order of smallness, yields

$$1 + i\,\varepsilon_1 + i\,\varepsilon_4 = 1$$

or

$$\varepsilon_4 = -\varepsilon_1$$

By virtue of this relation, Equation (3.47) takes the form

$$I + B = \begin{bmatrix} 1 + i\varepsilon_1 & \varepsilon_2 + i\varepsilon_3 \\ -\varepsilon_2 + i\varepsilon_3 & 1 - i\varepsilon_1 \end{bmatrix} \tag{3.48}$$

The matrix on the right-hand side of Equation (3.48) shows that SU(2) is a three-parameter group. Now, in view of Equation (3.48), Equation (3.44) for the infinitesimal transformation $x' = (I + B)x$ gives

$$\begin{bmatrix} x_1' \\ x_2' \end{bmatrix} = \begin{bmatrix} 1+i\varepsilon_1 & \varepsilon_2 + i\varepsilon_3 \\ -\varepsilon_2 + i\varepsilon_3 & 1-i\varepsilon_1 \end{bmatrix} \begin{bmatrix} x_1 \\ x_2 \end{bmatrix}$$

Comparing the corresponding elements on the two sides of this matrix equation, we get

$$x_1' = (1 + i\,\varepsilon_1)x_1 + (\varepsilon_2 + i\,\varepsilon_3)x_2$$

and

$$x_2' = (-\varepsilon_2 + i\,\varepsilon_3)x_1 + (1 - i\,\varepsilon_1)x_2$$

or

$$x_1' - x_1 = i\varepsilon_1 x_1 + \varepsilon_2 x_2 + i\varepsilon_3 x_2 \tag{3.49a}$$

and

$$x_2' - x_2 = -\,i\varepsilon_1 x_2 - \varepsilon_2 x_1 + i\varepsilon_3 x_1 \tag{3.49b}$$

But from Equation (3.25), we have

$$x_i' - x_i = \varepsilon_k u_{ik} = \varepsilon_1 u_{i1} + \varepsilon_2 u_{i2} + \varepsilon_3 u_{i3}, \quad i = 1, 2$$

For $i = 1$ and $i = 2$, Equation (3.25′) yields

$$x_1' - x_1 = \varepsilon_1 u_{11} + \varepsilon_2 u_{12} + \varepsilon_3 u_{13} \tag{3.50a}$$

$$x_2' - x_2 = \varepsilon_1 u_{21} + \varepsilon_2 u_{22} + \varepsilon_3 u_{23} \tag{3.50b}$$

Comparing Equations (3.49) and (3.50), we get

$$u_{11} = ix_1,\ u_{12} = x_2,\ u_{13} = ix_2$$

$$u_{21} = -ix_2,\ u_{22} = -x_1,\ u_{23} = ix_1$$

The generators L_k are given by

$$L_k = u_{ik}\frac{\partial}{\partial x_i} = u_{1k}\frac{\partial}{\partial x_1} + u_{2k}\frac{\partial}{\partial x_2}, \qquad k = 1, 2, 3$$

where the summation over i is from 1 to 2, as there are two variables. Substituting the expressions for u's, we get

$$L_1 = u_{11}\frac{\partial}{\partial x_1} + u_{21}\frac{\partial}{\partial x_2} = ix_1\frac{\partial}{\partial x_1}$$

$$L_2 = u_{12}\frac{\partial}{\partial x_1} + u_{22}\frac{\partial}{\partial x_2} = x_2\frac{\partial}{\partial x_1} - x_1\frac{\partial}{\partial x_2}$$

$$L_3 = u_{13}\frac{\partial}{\partial x_1} + u_{23}\frac{\partial}{\partial x_2} = ix_2\frac{\partial}{\partial x_1} + ix_1\frac{\partial}{\partial x_2}$$

If we use x for x_1 and y for x_2, we can write the generators as

$$L_1 = i\left(x\frac{\partial}{\partial x} - y\frac{\partial}{\partial y}\right)$$

$$L_2 = y\frac{\partial}{\partial x} - x\frac{\partial}{\partial y} \tag{3.51}$$

$$L_3 = i\left(y\frac{\partial}{\partial x} + x\frac{\partial}{\partial y}\right)$$

PROBLEM 3.18

Find the commutation relations satisfied by the generators of this group. Hence, show that the commutator of any two generators of the group is a linear combination of its generators.

3.8 Generators of SU(3)

The group SU(3), the special unitary group in a complex space of three dimensions, is characterized by transformations of the type

$$x' = A\,x$$

where the 3 × 3 transformation matrix

$$A = \begin{bmatrix} a_{11} & a_{12} & a_{13} \\ a_{21} & a_{22} & a_{23} \\ a_{31} & a_{32} & a_{33} \end{bmatrix}$$

satisfies the conditions

$$A^{\dagger}A = I \qquad \text{and} \qquad |A| = 1$$

The vectors x and x′ are given by

$$x = \begin{bmatrix} x_1 \\ \\ x_2 \end{bmatrix} \quad \text{and} \quad x' = \begin{bmatrix} x_1' \\ \\ x_2' \end{bmatrix}$$

The identity transformation, $x' = Ix = x$, is obtained by putting $A = I$. The infinitesimal transformation is obtained by choosing $A = I + B$:

$$x' = (I + B)x \tag{3.52}$$

where the elements of 3 x 3 matrix B are infinitesimal. Because $(I + B)$ is the transformation matrix for the infinitesimal transformation and as, in SU(3), every such matrix must be unitary, we have

$$(I + B)^{\dagger}(I + B) = I$$

To the first order of smallness, this yields

$$B^{\dagger} = -B$$

Let us write

$$B = \begin{bmatrix} a & b & c \\ d & e & f \\ g & h & j \end{bmatrix} \tag{3.53}$$

Then the condition $B^{\dagger} = -B$ gives

$$a = -a^*, b = -d^*, c = -g^*, e = -e^*, f = -h^*, j = -j^*$$

These relations show that the diagonal elements of the matrix B are pure imaginary, and for $j \neq i$ the ij-element is related to the ji-element such that if ij-element is $\alpha + i\beta$, then ji-element is $-\alpha + i\beta$. In terms of infinitesimal parameters, the elements of the matrix B may be written as

$$a = i\,\varepsilon_1, b = \varepsilon_2 + i\varepsilon_3, c = \varepsilon_4 + i\varepsilon_5, d = -\varepsilon_2 + i\varepsilon_3$$

$$e = i\varepsilon_6, f = \varepsilon_7 + i\varepsilon_8, g = -\varepsilon_4 + i\varepsilon_5, h = -\varepsilon_7 + i\varepsilon_8, \text{and } j = i\varepsilon_9$$

Moreover, since, by definition, the transformation matrices are also unimodular, we should have

$$|I + B| = 1 \tag{3.54}$$

To first order of smallness, Equation (3.54) gives

$$\varepsilon_1 + \varepsilon_6 + \varepsilon_9 = 0$$

so that B takes the form

$$B = \begin{bmatrix} i\varepsilon_1 & \varepsilon_2 + i\varepsilon_3 & \varepsilon_4 + i\varepsilon_5 \\ -\varepsilon_2 + i\varepsilon_3 & i\varepsilon_6 & \varepsilon_7 + i\varepsilon_8 \\ -\varepsilon_4 + i\varepsilon_5 & -\varepsilon_7 + i\varepsilon_8 & -i(\varepsilon_1 + \varepsilon_6) \end{bmatrix} \tag{3.55}$$

Equation (3.55) shows that SU(3) is an eight-parameter group. This is as it must be because the order of SU(3) is $3^2 - 1 = 8$. Substituting this expression for B in Equation (3.52) and comparing the coefficients of ε_k thus obtained with those in Equation (3.25), viz.

$$x_i' - x_i = \varepsilon_k u_{ik}, \quad i = 1, 2, 3 \tag{3.25'}$$

we get

$$u_{11} = u_{23} = u_{35} = ix_1$$

$$u_{12} = -u_{37} = x_2$$

$$u_{13} = u_{26} = u_{38} = ix_2$$

$$u_{14} = u_{27} = x_3$$

$$u_{15} = u_{28} = -u_{31} = -u_{36} = ix_3$$

$$u_{21} = u_{16} = u_{17} = u_{18} = u_{24} = u_{25} = u_{32} = u_{33} = 0$$

$$u_{22} = u_{34} = -x_1$$

By making use of Formula (3.28) for the generators, viz.

$$L_k = u_{ik} \frac{\partial}{\partial x_i}$$

we get

$$L_k = u_{1k} \frac{\partial}{\partial x_1} + u_{2k} \frac{\partial}{\partial x_2} + u_{3k} \frac{\partial}{\partial x_3}$$

Substituting the values of u_{ik}, we obtain

$$L_1 = i\left(x_1\frac{\partial}{\partial x_1} - x_3\frac{\partial}{\partial x_3}\right) \qquad L_2 = x_2\frac{\partial}{\partial x_1} - x_1\frac{\partial}{\partial x_2}$$

$$L_3 = i\left(x_2\frac{\partial}{\partial x_1} + x_1\frac{\partial}{\partial x_2}\right) \qquad L_4 = x_3\frac{\partial}{\partial x_1} - x_1\frac{\partial}{\partial x_3}$$

$$L_5 = i\left(x_3\frac{\partial}{\partial x_1} + x_1\frac{\partial}{\partial x_3}\right) \qquad L_6 = i\left(x_2\frac{\partial}{\partial x_2} - x_3\frac{\partial}{\partial x_3}\right)$$

$$L_7 = x_3\frac{\partial}{\partial x_2} - x_2\frac{\partial}{\partial x_3} \qquad L_8 = i\left(x_3\frac{\partial}{\partial x_2} - x_2\frac{\partial}{\partial x_3}\right)$$

If we use x for x_1, y for x_2 and z for x_3, we can write the eight generators as

$$L_1 = i\left(x\frac{\partial}{\partial x} - z\frac{\partial}{\partial z}\right) \qquad L_2 = \left(y\frac{\partial}{\partial x} - x\frac{\partial}{\partial y}\right)$$

$$L_3 = i\left(y\frac{\partial}{\partial x} + x\frac{\partial}{\partial y}\right) \qquad L_4 = z\frac{\partial}{\partial x} - x\frac{\partial}{\partial z}$$

$$L_5 = i\left(z\frac{\partial}{\partial x} + x\frac{\partial}{\partial z}\right) \qquad L_6 = i\left(y\frac{\partial}{\partial y} - z\frac{\partial}{\partial z}\right)$$

$$L_7 = z\frac{\partial}{\partial y} - y\frac{\partial}{\partial z} \qquad L_8 = i\left(z\frac{\partial}{\partial y} + y\frac{\partial}{\partial z}\right)$$

PROBLEM 3.19

Show that the total number of commutation relations for SU(n) is $r(r-1)/2$, where r is the number of generators of the group.

PROBLEM 3.20

Find the generators of the group SO(4).

3.9 Generators and Parameterization of a Group

Consider the group of inhomogeneous linear transformations

$$x' = ax + b, \qquad a \neq 0$$

We have shown that this two-parameter group of transformations has two generators L_1 and L_2 given by

$$L_1 = x\frac{d}{dx} \quad \text{and} \quad L_2 = \frac{d}{dx}$$

However, this group can also be parameterized as

$$x' = (a_1 + a_2)\,x + 1 + (a_1 - a_2), \qquad (a_1 + a_2) \neq 0$$

If we calculate the generators X_1 and X_2 by writing the transformation equation in the above form, we get

$$X_1 = (x+1)\frac{d}{dx} \quad \text{and} \quad X_2 = (x-1)\frac{d}{dx}$$

This example illustrates the following points, which are true in general:

1. Generators of a Lie group are not unique. The expressions for them depend upon the way the group is parameterized.
2. Generators of a Lie group with a particular parameterization are linearly independent, and their number is equal to that of essential parameters.
3. Generators obtained by different parameterizations are linear combinations of each other.

Thus, in the previous example, we have

$$X_1 = L_1 + L_2$$

$$X_2 = L_1 - L_2$$

and

$$L_1 = \frac{(X_1 + X_2)}{2} \quad L_2 = \frac{(X_1 - X_2)}{2}$$

It can also be proved that with a suitable parameterization, all the generators of a Lie group can be made Hermitian.

From a set of Hermitian generators L_k, we can obtain a representation of the group by means of unitary operators $U(\varepsilon_k)$, which are given by

$$U(\varepsilon_k) = \exp(i\varepsilon_k L_k) \equiv U_k$$

where ε_k are real parameters that vary continuously from the identity. An element of the group that can be reached continuously from the identity can be written as a product of these operators.

3.10 Matrix Representatives of Generators

We will now describe a method for finding the matrices representing the generators of a Lie group. We shall illustrate the technique by reference to two examples but omit the proof.

Consider the (real) group O(2). We have seen that in this case the matrix B is given by

$$B = \begin{bmatrix} 0 & -\varepsilon_1 \\ \varepsilon_1 & 0 \end{bmatrix} \tag{3.43a'}$$

Differentiating with respect to ε_1, we get

$$L = \begin{bmatrix} 0 & -1 \\ 1 & 0 \end{bmatrix}$$

This is the generator of O(2).

Let us next consider SU(2). Here

$$B = \begin{bmatrix} i\varepsilon_1 & \varepsilon_2 + i\varepsilon_3 \\ -\varepsilon_2 + i\varepsilon_3 & -i\varepsilon_1 \end{bmatrix}$$

Differentiating in turn with respect to ε_1, ε_2, and ε_3, we get

$$L_1 = \begin{bmatrix} 1 & 0 \\ 0 & -1 \end{bmatrix}, \quad L_2 = \begin{bmatrix} 0 & 1 \\ -1 & 0 \end{bmatrix}, \quad L_3 = \begin{bmatrix} 0 & 1 \\ 1 & 0 \end{bmatrix}$$

PROBLEM 3.21

Describe the formula for determining the matrix representatives of the generators of a Lie group of nonsingular linear transformations

$$x' = Ax, \qquad |A| \neq 0$$

PROBLEM 3.22

Calculate the matrix representatives of the generators of U(2) and SU(3).

PROBLEM 3.23

Show that the matrix representatives of the generators of SU(2) can be expressed as traceless Hermitian matrices.

3.11 Structure Constants

It has already been shown with the help of a few examples and can be proved in general that the commutator of any two generators of a Lie group is a linear combination of its generators. Consider an r-parameter semisimple Lie group. Let $L_1, L_2, \ldots, L_r$ be r generators of this group. Let L_A and L_B be any two of these generators. Then the commutator of L_A and L_B is given by

$$[L_A, L_B] = \sum_D C^D_{AB} L_D$$

The constants $C^D{}_{AB}$ are in general complex numbers and are called *structure constants*. The notation for structure constants is such that their lower indices reflect the generators whose commutator is being considered, whereas the upper index refers to the generator whose coefficient the structure constant is. By using the summation convention, the previous equation can be written as

$$[L_A, L_B] = C^D{}_{AB} L_D \tag{3.56}$$

The following two properties of structure constants can be easily established:

1. Interchanging A and B throughout Equation (3.56), we get

   $$[L_B, L_A] = C^D{}_{BA} L_D$$

   However, changing the order of the two generators changes the sign of the commutator. Therefore, we get

   $$[L_A, L_B] = -[L_B, L_A] = -C^D{}_{BA} L_D \tag{3.57}$$

   Subtracting Equation (3.57) from Equation (3.56), we get

   $$(C^D{}_{AB} + C^D{}_{BA}) L_D = 0$$

   Since the generators $L_1, L_2, \ldots, L_r$ are linearly independent, this equation can hold only if the coefficient of each L_D is equal to zero, that is, if

   $$C^D{}_{AB} = -C^D{}_{BA} \tag{3.58}$$

   This equation shows that the structure constants are antisymmetric in their lower indices: interchanging their lower indices, changes their sign.

2. If P, Q, and R are any three operators, then, according to the Jacobi identity, these satisfy the relation

   $$[P, [Q, R]] + [Q, [R, P]] + [R, [P, Q]] \equiv 0$$

By considering the generators L_A, L_B, L_D of any Lie group and making use of this identity, we get

$$[L_A, [L_B, L_D]] + [L_B, [L_D, L_A]] + [L_D, [L_A, L_B]] \equiv 0$$

Since

$$[L_B, L_D] = C^E{}_{BD} L_E, \text{ etc.}$$

the previous identity yields

$$[L_A, C^E{}_{BD} L_E] + [L_B, C^E{}_{DA} L_E] + [L_D, C^E{}_{AB} L_E] = 0$$

or

$$C^E{}_{BD} [L_A, L_E] + C^E{}_{DA} [L_B, L_E] + C^E{}_{AB} [L_D, L_E] = 0$$

or

$$C^E{}_{BD} C^F{}_{AE} L_F + C^E{}_{DA} C^F{}_{BE} L_F + C^E{}_{AB} C^F{}_{DE} L_F = 0$$

or

$$(C^E{}_{BD} C^F{}_{AE} + C^E{}_{DA} C^F{}_{BE} + C^E{}_{AB} C^F{}_{DE}) L_F = 0$$

Since the generators L_F are linearly independent, the coefficient of each one of them must be zero. Therefore, we must have

$$C^E{}_{BD} C^F{}_{AE} + C^E{}_{DA} C^F{}_{BE} + C^E{}_{AB} C^F{}_{DE} = 0 \tag{3.59}$$

Thus conditions (3.58) and (3.59) are necessary for the existence of generators satisfying Equation (3.56).

We will now show that they are also *sufficient*, that is, whenever these two conditions are satisfied, there exist operators that satisfy Equation (3.56). We proceed in the following manner.

It is given that the structure constants satisfy Relation (3.59), namely,

$$C^E{}_{BD} C^F{}_{AE} + C^E{}_{DA} C^F{}_{BE} + C^E{}_{AB} C^F{}_{DE} = 0$$

Since C's are given to be antisymmetric in their lower indices, this equation may be written as

$$C^E{}_{BD} C^F{}_{AE} - C^E{}_{AD} C^F{}_{BE} - C^E{}_{AB} C^F{}_{ED} = 0$$

REMARK

We know from matrix theory that

$$(AB)_{ij} = A_{ik}B_{kj}$$

where $(AB)_{ij}$ stands for the element in the i-th row and j-th column of the product matrix AB. The notation is slightly changed and we write the element in the i-th row and j-th column of AB as $(AB)_i^{\ j}$. Thus

$$(AB)_i^{\ j} = A_i^{\ k}B_k^{\ j}$$

This equation may now be written as

$$(C_B)^E{}_D(C_A)^F{}_E - (C_A)^E{}_D(C_B)^F{}_E = C^E{}_{AB}(C_E)^F{}_D$$

or

$$(C_A)^E{}_D(C_B)^F{}_E - (C_B)^E{}_D(C_A)^F{}_E = C^E{}_{AB}(-C_E)^F{}_D$$

or

$$(C_AC_B)^F{}_D - (C_BC_A)^F{}_D = C^E{}_{AB}(-C_E)^F{}_D$$

This is merely the DF-element of the matrix equation

$$C_AC_B - C_BC_A = C^E{}_{AB}(-C_E)$$

Since DF is arbitrary, so that the equation for the elements may be valid, the matrix equation must hold. But this matrix equation can be written as

$$[-C_A, -C_B] = C^E{}_{AB}(-C_E) \tag{3.60}$$

This shows that Equations (3.58) and (3.59) are sufficient for the existence of matrices satisfying Relation (3.56).

The matrices $-C_A$, $-C_B$, …, $-C_R$ form an explicit representation of the generators L_A, L_B, …, L_R of the group. This representation is called *regular* or *adjoint representation* of the group.

3.12 Rank of a Lie Group

The rank ℓ of a Lie group is determined by the maximal number of its linearly independent generators that mutually commute. If none of the generators of a group commutes with any other generator of the group, the rank is said to be 1.

Example 3.4

The rank of the group O(3) is 1 as its generators do not commute.

It may be remarked that in various linearly independent sets of generators of a Lie group obtained by using different parameterizations, although the total number of generators in each set remains the same, it is not essential that the number of commuting generators of the group should also be the same. For instance, if there are three linearly independent generators L_1, L_2, L_3 of a group such that $[L_1, L_2] = 0$, $[L_1, L_3] \neq 0$, $[L_2, L_3] \neq 0$—that is, if it contains two commuting operators—then the set $\{L_1 + L_3, L_2 + L_3, L_3\}$ also contains three linearly independent generators but does not contain any commuting pair. Hence, for determining the rank of a group, we must consider its generators along with all possible linear combinations and then find the maximum number of generators that mutually commute.

PROBLEM 3.24

Show that for an Abelian Lie group, all the structure constants are zeros.

3.13 Lie Algebras

If for the elements x_1, x_2, ... of a linear vector space V over a field F, an additional binary operation of multiplication can also be defined such that

(i) The product of any two elements of the vector space is also a member of the vector space, that is, for all $x_1, x_2 \in V$, the product $x_1, x_2 \in V$

(ii) For any $x_1, x_2 \in V$ and $a \in F$, we have $ax_1 x_2 = x_1 a x_2$

then the elements x_1, x_2, ... are said to form a linear algebra. The linear algebra is said to be real or complex since F is the field of real or complex numbers. Now consider the set of all linear combinations α, β,... of the generators of a Lie group G. This set forms a vector space, say V, with the generators of the group as basis vectors. We know that the commutator of any two generators of a Lie group is a linear combination of the generators of the group. Therefore, if commutation is taken as the additional binary operation of multiplication, then for $\alpha, \beta \in V$, the commutator $[\alpha, \beta]$ of any two linear combinations of the generators is also a linear combination of the generators and is hence an element of the vector space spanned by the generators of the Lie group. This vector space therefore forms a linear algebra, as, under two binary operations, addition and commutation, the elements of the vector space yield elements of the same vector space. In this specific case, the linear algebra is called a *Lie algebra*.

The significance of Lie algebras stems from the fact that Lie groups, which are locally isomorphic (i.e., isomorphic in a neighborhood of the identity), have the same Lie algebra. Therefore, by considering a single Lie algebra, one is, in effect, considering a whole class of Lie groups, all of which are locally isomorphic in the sense that neighborhoods of the identity element exist that are isomorphic. Hence, if we are interested only in local properties of Lie groups—that is, properties determined by the behavior of the group in the neighborhood of the identity—it is convenient to study Lie algebras rather than Lie groups. However, it may be emphasized that the Lie algebra of a group does not tell about the global properties of the group.

The aforementioned correspondence between Lie groups and their Lie algebras makes it possible to go freely from Lie groups to Lie algebras so that the definitions and local properties of Lie groups may be easily extended to the corresponding Lie algebras.

It may be pointed out that the Lie algebra of a direct product of two groups is just the algebra of the generators of both. Moreover, the Lie algebra of the direct product of two simple groups is semisimple. Since, in high energy physics, we will be mainly concerned with semisimple compact Lie groups, we shall confine our discussion to such groups.

3.14 Commutation Relations between the Generators of a Semisimple Lie Group

Consider an r-parameter semisimple Lie group of rank ℓ. It will have r generators $L_1, L_2, \ldots, L_r$. The maximum number of its mutually commuting generators will be denoted by ℓ and written as $H_1, H_2, \ldots, H_\ell$. The remaining $r - \ell$ generators will be denoted by $E_1, E_2, \ldots, E_{r-\ell}$. We will use the Latin indices i, j, k, etc., for the generators of type H and Greek indices α, β, γ, etc., for the generators of type E. The indices A, B, C, ... will be used to refer to the complete set of generators. Thus, L_A stands for any one of the complete set of generators, H_i for any one of the mutually commuting set of generators, and E_α for a generator belonging to the remaining set of generators. We shall also assume that the parameterization has been done in such a way that all the generators are Hermitian. Let us now consider the commutation relations between H_i and L_A. Since, by definition, the generators of type H commute with each other, we must have

$$[H_i, H_j] = 0 = C^D{}_{ij} L_D \tag{3.61a}$$

as the commutator of any two generators of a group is a linear combination of the generators of that group. Since the generators L_D are linearly independent, we must have $C^D{}_{ij} = 0$ for all values of i, j, and D. Moreover, we can write

$$[H_i, E_\alpha] = C^D{}_{i\alpha} L_D \tag{3.61b}$$

It is possible to parameterize the group in such a manner that $C^D_{i\alpha} = 0$ for all values of $D \neq \alpha$. Then, on the right-hand side Equation (3.61b), only one term is left for which $D = \alpha$. This yields

$$\begin{aligned}[H_i, E_\alpha] &= C^\alpha_{i\alpha}\, L_\alpha \text{ (no summation over } \alpha) \\ &= r_i(\alpha)\, E_\alpha \end{aligned} \tag{3.62}$$

where we have written $r_i(\alpha)$ for $C^\alpha_{i\alpha}$. By virtue of Equation (3.62), for $i = 1, 2, \ldots, \ell$, we have

$$\begin{aligned} [H_1, E_\alpha] &= r_1(\alpha)\, E_\alpha \\ [H_2, E_\alpha] &= r_2(\alpha)\, E_\alpha \\ &\cdots\cdots\cdots \\ [H_\ell, E_\alpha] &= r_\ell(\alpha)\, E_\alpha \end{aligned} \tag{3.63}$$

Thus, for a given generator E_α, by finding the commutators of H_i and E_α, $i = 1, 2, \ldots, \ell$, we obtain ℓ numbers $r_1(\alpha), r_2(\alpha), \ldots, r_\ell(\alpha)$. These numbers may be considered as the components of a vector in an ℓ-dimensional space. This vector is known as the *root vector* or just the *root* and is denoted by $\mathbf{r}(\alpha)$ or $\boldsymbol{\alpha}$:

$$\boldsymbol{\alpha} \equiv \mathbf{r}(\alpha) = (r_1(\alpha), r_2(\alpha), \ldots, r_\ell(\alpha)) \tag{3.64}$$

The ℓ-dimensional space is called the *root space*. For every E-type generator, we have a root vector. Since such generators are $(r - \ell)$ in number, a semi-simple Lie group will have $(r - \ell)$ root vectors.

Notice that all the components of a root vector $\boldsymbol{\alpha}$ cannot be zero because then the corresponding generator E_α will be commuting with all H_i and therefore, against the initial assumption, will not be an E-type generator: it should then be an H-type generator, which is not true.

PROBLEM 3.25

Can an E-type generator commute with all H-type generators?

PROBLEM 3.26

Can any one of the numbers $r_i(\alpha)$ be zero? If so, what is the maximum number of $r_i(\alpha)$ that can be zero?

Moreover, as $[H_i, H_j] = 0$, the corresponding root vectors, each one obtained by choosing a fixed H_j and varying H_i, $i = 1, 2, \ldots, \ell$, will be null vectors. The number of zero root vectors is evidently equal to the rank ℓ of the group.

Hence the total number of roots is r, the same as the number of essential parameters or the maximum number of linearly independent generators.

It can be proved that if $\mathbf{r}(\alpha) \equiv \boldsymbol{\alpha}$ is a root of a semisimple Lie group, then $\mathbf{r}(-\alpha) \equiv -\boldsymbol{\alpha}$ is also a root of that group. The corresponding E-type generator will be denoted by $E_{-\alpha}$. Further, it can be proved (see Appendix A) that

(1) If $\boldsymbol{\alpha}$ and $\boldsymbol{\beta}$ are two roots such that $\boldsymbol{\beta} = -\boldsymbol{\alpha}$, then

$$[E_\alpha, E_{-\alpha}] = C^i_{\alpha,-\alpha} H_i = r^i(\alpha)H_i = r_i(\alpha)H_i \tag{3.65}$$

$$\text{where } r^i(\alpha) = r_i(\alpha) = C^i_{\alpha,-\alpha} \tag{3.66}$$

(2) If $\boldsymbol{\alpha}$ and $\boldsymbol{\beta}$ are two roots, then

$$\begin{aligned}[E_\alpha, E_\beta] &= C^D{}_{\alpha\beta}L_D = C^{\alpha+\beta}{}_{\alpha\beta}L_{\alpha+\beta} \\ &= N_{\alpha\beta}\, E_{\alpha+\beta}, \text{ if } \boldsymbol{\alpha} + \boldsymbol{\beta} \text{ is a nonvanishing root} \\ &= 0, \text{ if } \boldsymbol{\alpha} + \boldsymbol{\beta} \text{ is not a root}\end{aligned} \tag{3.67}$$

where $N_{\alpha\beta}$ is a nonzero structure constant.

The components $r_i(\alpha)$ also satisfy the orthonormality condition

$$\sum_\alpha r_i(\alpha) r_j(\alpha) = \delta_{ij} \tag{3.68}$$

where α can have $r - \ell$ values.

To sum up, the generators of a semisimple Lie group satisfy the following commutation relations:

$$[H_i, H_j] = 0 \tag{3.69a}$$

$$[H_i, E_\alpha] = r_i(\alpha)\, E_\alpha \tag{3.69b}$$

$$[E_\alpha, E_{-\alpha}] = r_i(\alpha)\, H_i \tag{3.69c}$$

$$\begin{aligned}[E_\alpha, E_\beta] &= N_{\alpha\beta}\, E_{\alpha+\beta}, \text{ if } \boldsymbol{\alpha} + \boldsymbol{\beta} \text{ is a nonvanishing root} \\ &= 0, \text{ if } \boldsymbol{\alpha} + \boldsymbol{\beta} \text{ is not a root}\end{aligned} \tag{3.69d}$$

where structure constant $N_{\alpha\beta} \neq 0$, and

$$\sum_\alpha r_i(\alpha) r_j(\alpha) = \delta_{ij} \tag{3.69e}$$

These commutation relations show that if H_i and E_α are known, the structure constants $N_{\alpha\beta}$ and $r_i(\alpha)$ can be determined. These equations constitute what is known as the standard form of the commutation relations. If we consider the operators H_i as the components of a vector **operator H**, we can write the commutation relations (3.69b) and (3.69c) as

$$[\mathbf{H}, E_\alpha] = \mathbf{r}(\alpha)E_\alpha \tag{3.70}$$

and

$$[E_\alpha, E_{-\alpha}] = \mathbf{r}\cdot\mathbf{H} \tag{3.71}$$

Next, we define positive and simple roots.

A root of a group is said to be *positive* if its first nonzero component is positive. A positive root of a group is said to be *simple* if it cannot be decomposed into a sum of two positive roots of the same group.

3.15 Properties of the Roots

We shall now state some important properties of the roots of a semisimple Lie group and give proofs in most of the cases.

1. The nonvanishing roots $\mathbf{r}(\alpha) \equiv \boldsymbol{\alpha}$ are nondegenerate.
2. The roots $\mathbf{r}(\alpha) \equiv \boldsymbol{\alpha}$ are the differences of the eigenvalues of **H**. This actually expresses the fact that the components of a root are the differences of the eigenvalues of H_i.

PROOF

Let $|m_i>$ be an eigenvector of H_i corresponding to the eigenvalue m_i. Then we can write

$$H_i|m_i> = m_i|m_i> \tag{3.72}$$

We have already seen that

$$[H_i, E_\alpha] = r_i(\alpha)E_\alpha \tag{3.69b$'$}$$

Operating upon the vector $|m_i>$, we obtain

$$[H_i, E_\alpha]|m_i> = r_i(\alpha)\, E_\alpha|m_i>$$

or

$$H_iE_\alpha|m_i> - E_\alpha H_i|m_i> = r_i(\alpha)\, E_\alpha|m_i>$$

Substituting the expression for $H_i|m_i>$ from Equation (3.72) in the previous equation, we get

$$H_iE_\alpha|m_i> - m_iE_\alpha|m_i> = r_i(\alpha)\, E_\alpha|m_i>$$

or

$$H_i(E_\alpha|m_i>) = [m_i + r_i(\alpha)](E_\alpha|m_i>) \tag{3.73}$$

This equation shows that $[m_i + r_i(\alpha)]$ is also an eigenvalue of the generator H_i (but corresponds to a different eigenvector, viz. $E_\alpha|m_i>$). Comparing Equations (3.72) and (3.73), we note that $r_i(\alpha)$, the i-th component of the root vector $\mathbf{r}(\alpha)$, is the difference of the eigenvalues m_i and $[m_i + r_i(\alpha)]$ of the Hermitian operator H_i corresponding to different eigenvectors $|m_i>$ and $E_\alpha|m_i>$. This fact is also expressed by stating that the root $\mathbf{r}(\alpha) \equiv \boldsymbol{\alpha}$ is the difference of the eigenvalues of $\mathbf{H}$. ■

3. The roots are real.

PROOF

Since the generators H_i are Hermitian, their eigenvalues must be real. Therefore, the components of a root, being the differences of the eigenvalues of H_i, must also be real. Hence, the roots are real. ■

4. The total number of nonzero distinct roots of a semisimple Lie group of order r and rank ℓ is $(r - \ell)/2$.

PROOF

We have seen that a semisimple Lie group of order r and rank ℓ has $r - \ell$ nonzero roots. However, if $\mathbf{r}(\alpha)$ is a root, then $-\mathbf{r}(\alpha) \equiv -\alpha$ is also a root. This reduces the number of nonzero distinct roots to $(r-\ell)/2$. ■

5. If $\boldsymbol{\alpha}$ and $\boldsymbol{\beta}$ are two nonvanishing roots, then $2\,\boldsymbol{\alpha}\cdot\boldsymbol{\beta}/\boldsymbol{\beta}\cdot\boldsymbol{\beta} \equiv g(\alpha/\beta)$, say) is an integer, called a *Cartan integer.*

PROOF

See Appendix B. ■

6. If $\boldsymbol{\alpha}$ and $\boldsymbol{\beta}$ are two nonzero roots, then $\boldsymbol{\alpha} - g(\alpha/\beta)\boldsymbol{\beta}$ is also a nonzero root.

PROOF

See Appendix B. ■

7. If $\boldsymbol{\alpha}$ and $c\boldsymbol{\alpha}$ are two nonvanishing roots of a semisimple Lie group, then $c = \pm 1$.

PROOF

If $\boldsymbol{\alpha}$ and $c\boldsymbol{\alpha}$ are two nonzero roots of a semisimple group, then, by virtue of property 5, we have:

$$2\,\frac{\alpha \cdot c\alpha}{c\alpha \cdot c\alpha} = a$$

where a is an integer. Since $\boldsymbol{\alpha}$ and c cannot vanish, the previous equation gives

$$c = \frac{2}{a} \tag{3.74}$$

Smilarly, taking $c\boldsymbol{\alpha}$ and $\boldsymbol{\alpha}$ as two nonzero roots, we have

$$2\frac{c\alpha \cdot \alpha}{\alpha \cdot \alpha} = b$$

where b is an integer. This yields

$$c = \frac{b}{2} \tag{3.75}$$

Eliminating c between Equations (3.74) and (3.75), we obtain

$$ab = 4 \tag{3.76}$$

Since a and b are both integers, the possible solutions of Equation (3.76) are

$$a = 1, b = 4; a = 2, b = 2; a = 4, b = 1$$
$$a = -1, b = -4; a = -2, b = -2; a = -4, b = -1$$

The corresponding values of c are obtained by using either Equation (3.74) or Equation (3.75). These are

$$c = \pm 2,\ \pm\ 1,\ \pm\frac{1}{2}$$

If $\boldsymbol{\beta} = \boldsymbol{\alpha}$, then Equation (3.69d) gives

$$[E_\alpha, E_\alpha] = N_{\alpha\alpha}E_{2\alpha}, \text{ if } 2\,\boldsymbol{\alpha} \text{ is a nonvanishing root}$$
$$= 0, \text{ if } \boldsymbol{\alpha} + \boldsymbol{\beta} \text{ is not a root}$$

where $N_{\alpha\alpha}$ is nonzero. But an operator always commutes with itself. Therefore, we must have

$$[E_\alpha, E_\alpha] = 0$$

This means that $2\boldsymbol{\alpha}$ is not a root. This eliminates $c = 2$. The vector $-2\boldsymbol{\alpha}$ also cannot be a root because this would imply that $+2\boldsymbol{\alpha}$ is a root. This eliminates $c = -2$.

If $c = +1/2$, then $\boldsymbol{\alpha}$ and $\boldsymbol{\alpha}/2$ would be nonzero roots. However, by virtue of Equation (3.69d), by taking the two roots as $\boldsymbol{\alpha}/2$ and $\boldsymbol{\alpha}/2$, we have

$$[E_{\alpha/2}, E_{\alpha/2}] = N_{\alpha/2,\,\alpha/2}\, E_\alpha, \text{ if } \boldsymbol{\alpha} \text{ is a nonvanishing root}$$
$$= 0, \text{ if } \boldsymbol{\alpha} \text{ is not a root}$$

where $N_{\alpha/2,\alpha/2}$ is nonzero. But as

$$[E_{\alpha/2}, E_{\alpha/2}] = 0$$

this would imply that $\boldsymbol{\alpha}$ is not a root This is in contradiction with the initial assumption that $\boldsymbol{\alpha}$ is a root. Hence, $c \neq 1/2$ because it leads to a contradiction. Similarly, $c = -1/2$ is ruled out. Hence, c can be either +1 or −1. This proves the theorem. ■

8. The number of simple roots of a group is equal to the rank ℓ of the group.

3.16 Structure Constants $N_{\alpha\beta}$

The structure constants $N_{\alpha\beta}$ are given by (see Appendix C)

$$N_{\alpha\beta} = \pm\sqrt{j(k+1)\boldsymbol{\beta}\cdot\boldsymbol{\beta}/2} \tag{3.77a}$$

where $j > 0$ and $k \geq 0$ in the following chain of roots constructed from α and β:

$$\boldsymbol{\alpha} - k\boldsymbol{\beta}, \boldsymbol{\alpha} - (k-1)\boldsymbol{\beta}, \ldots, \boldsymbol{\alpha}, \ldots, \boldsymbol{\alpha} + j\boldsymbol{\beta}$$

The other structure constants can be determined from the following symmetry properties (see Appendix C)

$$N_{\alpha\beta} = -N_{\beta\alpha} = -N_{-\alpha,-\beta} = N_{\beta,-\alpha-\beta} = N_{-\alpha-\beta,\alpha} \tag{3.77b}$$

The signs of structure constants cannot be determined uniquely by Equation (3.77a), but they must be consistent with (3.77b). The choice of plus or minus sign, if made consistently, does not make any difference because consistent signs yield only isomorphic groups.

We now state two important theorems but omit their proofs.

Theorem 3.1

There is only a finite number of simple Lie algebras of any rank. ■

Theorem 3.2

Every semisimple compact group can be written as a direct product of simple compact groups. ■

Theorem 3.3 implies that there is only a finite number of semisimple compact algebras of a given rank. These theorems are helpful in the classification of groups.

3.17 Classification of Simple Groups

We know that if $\boldsymbol{\alpha}$ and $\boldsymbol{\beta}$ are two roots of a simple Lie group then

$$2\frac{\boldsymbol{\alpha}\cdot\boldsymbol{\beta}}{\boldsymbol{\beta}\cdot\boldsymbol{\beta}} = a \quad \text{an integer} \tag{3.78}$$

and

$$2\frac{\boldsymbol{\beta}\cdot\boldsymbol{\alpha}}{\boldsymbol{\alpha}\cdot\boldsymbol{\alpha}} = b \quad \text{an integer} \tag{3.79}$$

Since $\boldsymbol{\beta}\cdot\boldsymbol{\alpha} = \boldsymbol{\alpha}\cdot\boldsymbol{\beta}$, the multiplication of Equations (3.78) and (3.79) gives

$$4\frac{(\boldsymbol{\alpha}\cdot\boldsymbol{\beta})^2}{\alpha^2\beta^2} = ab$$

or

$$\cos^2 \varphi_{\alpha\beta} = \frac{ab}{4} \tag{3.80}$$

where $\varphi_{\alpha\beta}$ the angle between the vectors $\boldsymbol{\alpha}$ and $\boldsymbol{\beta}$.

Since a and b are integers, for $\varphi \leq 90°$, Equation (3.80) can hold only for the following values of $\varphi_{\alpha\beta}$:

$$\varphi_{\alpha\beta} = 0°, 30°, 45°, 60°, 90°$$

But the zero angle is dropped because then $\boldsymbol{\beta}$ is in the same direction as $\boldsymbol{\alpha}$, and therefore we do not get any new information by this choice. The analysis for $\varphi_{\alpha\beta}$ values is then reduced to

$$\varphi_{\alpha\beta} = 30°, 45°, 60°, 90° \tag{3.81}$$

We have not included the values of $\varphi_{\alpha\beta} > 90°$ because they do not give any additional information. For instance, we do not include, say along with 45°, the angles 135°, 225°, and 315° because if the angle $(\boldsymbol{\alpha}, \boldsymbol{\beta})$ between the roots $\boldsymbol{\alpha}$ and $\boldsymbol{\beta}$ is 45°, then the angles $(\boldsymbol{\beta}, -\boldsymbol{\alpha}) = 135°$, $(\boldsymbol{\alpha}, -\boldsymbol{\beta}) = 225°$, and $(-\boldsymbol{\beta}, -\boldsymbol{\alpha}) = 315°$ yield the same information. The negative values are excluded for the same reason.

It can be shown that for different values of $\varphi_{\alpha\beta}$ in Equation (3.81), we get different classes of the group. A complete list of the simple groups with their respective orders and ranks is given in the following tables.

Classical Groups

Group	Rank (ℓ)	Order (r)
$A_l \equiv SU(n) \equiv SU(\ell+1)$	$\ell = 1, 2, \ldots$	$\ell^2 + 2\ell$
B_ℓ	$\ell = 2, 3, \ldots$	$2\ell^2 + \ell$
C_ℓ	$\ell = 3, 4, \ldots$	$2\ell^2 + \ell$
D_ℓ	$\ell = 4, 5, \ldots$	$2\ell^2 - \ell$

Exceptional Groups

Group	Rank	Order
$G_2(14)$	2	14
$F_4(52)$	4	52
$E_6(78)$	6	78
$E_7(133)$	7	133
$E_8(248)$	8	248

3.18 Roots of SU(2)

We shall now determine the roots of SU(2). For that purpose, we notice that the order r of this group is 3. It has, therefore, three roots. Since no two generators of SU(2) mutually commute, the rank ℓ of SU(2) is one and therefore only one of its roots is zero. Evidently, the remaining two roots are nonzeros. Moreover, as whenever $\boldsymbol{\alpha}$ is a root, $-\boldsymbol{\alpha}$ is also a root, only one of the two nonzero roots is independent: there is only one distinct nonzero root. Furthermore, the rank ℓ of SU(2) being one, each root has one component only. Let α be a nonzero root of SU(2). It may be written as $\boldsymbol{\alpha} = (a) \equiv a$. The other root of SU(2) is $-\boldsymbol{\alpha} = -(a) \equiv -a$. The normalization condition gives

$$\sum_{\alpha} r_1^2(\alpha) = 1$$

or

$$r^2{}_i\,(\alpha) + r^2{}_i(-\alpha) = 1$$

For i = 1, we have

$$r^2{}_1\,(\alpha) + r^2{}_1(-\alpha) = 1$$

That is, the sum of the squares of the first component (which in this case is the only component) of every root is 1. Substituting the values of one-dimensional roots, we have

$$a^2 + (-a)^2 = 1$$

or

$$a = \pm 1/\sqrt{2}$$

Thus, two nonzero roots of SU(2) are $a = \pm 1/\sqrt{2}$. Diagrammatically, we can represent the three roots, $-1/\sqrt{2},\ 1/\sqrt{2}$, as shown in Figure 3.3. These roots

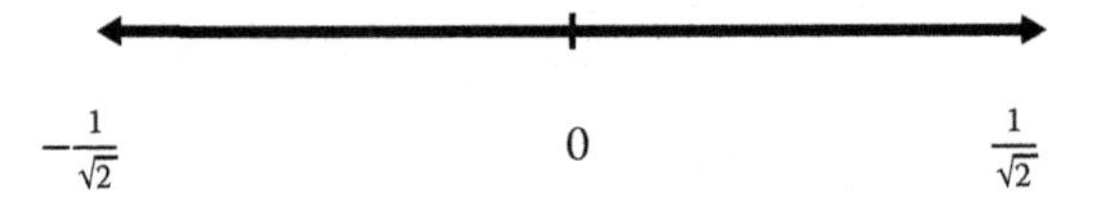

FIGURE 3.3
Root diagram for SU(2).

have been drawn from a common origin. This graphical representation of root vectors in an $\ell(=1)$-dimensional space is called a *root diagram.*

3.19 Roots of SU(3)

Let us next calculate all the roots of SU(3). Since this is a group of order $r = 8$ and rank $\ell = 2$, it will have $\ell = 2$ zero roots and three nonzero distinct roots, each root having $\ell = 2$ components. Let $\boldsymbol{\alpha}$, $\boldsymbol{\beta}$, and $\boldsymbol{\gamma}$ be three nonzero distinct roots of SU(3). Choosing one of the axes in a rectangular coordinate system in the two-dimensional root space along the root vector $\boldsymbol{\alpha}$, we may write

$$\boldsymbol{\alpha} = (a_1, 0) \tag{3.82a}$$

The other two roots may be written as

$$\boldsymbol{\beta} = (b_1, b_2) \tag{3.82b}$$

$$\boldsymbol{\gamma} = (c_1, c_2) \tag{3.82c}$$

To determine the values of a_1, b_1, b_2 and c_1, c_2, we proceed as follows.

Let $\boldsymbol{\alpha}$ and $\boldsymbol{\beta}$ be two adjacent roots of SU(3). For SU(3), the angle $\varphi_{\alpha\beta}$ between two adjacent roots is 60°. Now dividing Equation (3.78) by Equation (3.79) and taking the square root of both sides, we get

$$\frac{\alpha}{\beta} = \sqrt{\frac{a}{b}} \tag{3.83}$$

But for $\varphi_{\alpha\beta} = 60°$, Equation (3.80), viz., $\cos^2 \varphi_{\alpha\beta} = ab/4$, gives $ab = 1$. Since a and b are integers, this can hold only if $a = b = \pm 1$. Substituting any set of these values, viz., $a = b = 1$ or $a = b = -1$, in Equation (3.83), we obtain

$$\beta = \alpha \tag{3.84}$$

Thus, for SU(3), its adjacent roots and consequently all its roots have the same magnitude.

By virtue of property no. 6 of roots in Section 3.15, if $\boldsymbol{\alpha}$ and $\boldsymbol{\beta}$ are two roots, then the vector $\alpha - g(\alpha/\beta)\beta$ is also a root, where

$$g\left(\frac{\alpha}{\beta}\right) = 2\frac{\boldsymbol{\alpha} \cdot \boldsymbol{\beta}}{\boldsymbol{\beta} \cdot \boldsymbol{\beta}}$$

is an integer. Now

$$g\left(\frac{\alpha}{\beta}\right) = 2\frac{\boldsymbol{\alpha}\cdot\boldsymbol{\alpha}}{\boldsymbol{\beta}\cdot\boldsymbol{\beta}} = 2\frac{\alpha\beta\cos\varphi_{\alpha\beta}}{\beta^2} = 2\frac{\alpha\cos\varphi_{\alpha\beta}}{\beta}$$

But for adjacent roots of SU(3), say $\boldsymbol{\alpha}$ and $\boldsymbol{\beta}$, $\varphi_{\alpha\beta} = 60°$ and $\beta = \alpha$. Substituting these values of $\varphi_{\alpha\beta}$ and β in the previous equation, we get $g(\alpha/\beta) = 1$. Hence, $\boldsymbol{\alpha} - g(\alpha/\beta)\,\boldsymbol{\beta} = \boldsymbol{\alpha} - \boldsymbol{\beta}$ is also a root. This is the third root denoted by $\boldsymbol{\gamma}$. Since the roots $\boldsymbol{\alpha}$ and $\boldsymbol{\beta}$ are given as distinct (in fact, their magnitudes are equal but their directions are different), the third root $\boldsymbol{\gamma} = (c_1, c_2)$ of SU(3) is also distinct and is given by

$$\boldsymbol{\gamma} = \boldsymbol{\alpha} - \boldsymbol{\beta} = (a_1 - b_1, -b_2) \tag{3.85}$$

The constants c_1 and c_2 are therefore, respectively, equal to $a_1 - b_1$ and $-b_2$. Thus, the three distinct roots, $\boldsymbol{\alpha}$, $\boldsymbol{\beta}$, $\boldsymbol{\gamma}$, of SU(3) are

$$\begin{aligned}\boldsymbol{\alpha} &= (a_1, 0)\\ \boldsymbol{\beta} &= (b_1, b_2)\\ \boldsymbol{\gamma} &= (a_1 - b_1, -b_2)\end{aligned} \tag{3.86}$$

These distinct roots have been expressed in terms of three constants a_1, b_1, and b_2. Since whenever $\boldsymbol{\alpha}$ is a root, $-\boldsymbol{\alpha}$ is also a root, the remaining three non-zero roots are $-\boldsymbol{\alpha}$, $-\boldsymbol{\beta}$, and $-\boldsymbol{\gamma}$.

Let us now make use of the fact that the magnitudes of all the roots of SU(3) are equal: $\gamma = \beta = \alpha$ or $\gamma^2 = \beta^2 = \alpha^2$. By Equations (3.86), the relation $\beta^2 = \alpha^2$ yields

$$b_1^2 + b_2^2 = a_1^2$$

or

$$b_2 = \pm\sqrt{a_1^2 - b_1^2} \tag{3.87}$$

The components of the roots are such that, combined with the root property that if $\boldsymbol{\alpha}$ is a root, then $-\boldsymbol{\alpha}$ is also a root, the choice of the sign while taking the square root does not affect the final result. Choosing the positive sign and substituting the value of b_2 in Equations (3.86), we get

$$\begin{aligned}\boldsymbol{\alpha} &= (a_1, 0)\\ \boldsymbol{\beta} &= \left(b_1, \sqrt{a_1^2 - b_1^2}\right)\\ \boldsymbol{\gamma} &= \left(a_1 - b_1, -\sqrt{a_1^2 - b_1^2}\right)\end{aligned} \tag{3.88}$$

That is, the roots are expressed in terms of two constants a_1 and b_1.

To find the relation between a_1 and b_1, we again make use of the fact that the magnitudes of all the roots of SU(3) are equal. This time we consider the roots $\boldsymbol{\beta}$ and $\boldsymbol{\gamma}$. Then substituting the values of γ^2 and β^2 from Equation (3.88) in the relation $\gamma^2 = \beta^2$, we get

$$(a_1 - b_1)^2 + (a_1^2 - b_1^2) = b_1^2 + a_1^2 - b_1^2$$

or

$$(a_1 - b_1)^2 = b_1^2$$

or

$$a_1^2 + b_1^2 - 2a_1b_1 = b_1^2$$

or

$$a_1^2 - 2a_1b_1 = 0$$

or

$$a_1(a_1 - 2b_1) = 0$$

Therefore, either

$$a_1 = 0$$

or

$$a_1 = 2b_1$$

But $a_1 \neq 0$ because otherwise $\boldsymbol{\alpha}$ will be a zero root (0, 0). This is against the initial assumption that α is a nonzero root. Hence, we must have

$$a_1 = 2b_1$$

and consequently Equations (3.86) yield

$$\begin{aligned}
\boldsymbol{\alpha} &= (2b_1, 0) \\
\boldsymbol{\beta} &= \left(b_1,\ \sqrt{3}b_1\right) \\
\boldsymbol{\gamma} &= \left(b_1,\ -\sqrt{3}b_1\right)
\end{aligned} \tag{3.89}$$

The components of these three distinct roots have thus been expressed in terms of a single number b_1.

To determine the value of b_1, we make use of the normalization condition. The roots of a group are said to be orthonormalized if

$$\sum_{\alpha} r_i(\alpha) r_j(\alpha) = \delta_{ij} \tag{3.69e'}$$

For $j = i$, this condition gives

$$\sum_{\alpha} r_i^2(\alpha) = 1$$

That is, the sum of the squares of i-th components of all the vectors is unity. This yields

$$r_i^2(\alpha) + r_i^2(\beta) + r_i^2(\gamma) + r_i^2(-\alpha) + r_i^2(-\beta) + r_i^2(-\gamma) = 1$$

But as the i-th component of the root $\boldsymbol{\alpha}$ is just the negative of the i-th component of the root $-\boldsymbol{\alpha}$, the squares of their corresponding components must be equal: $r^2{}_i(\alpha) = r^2{}_i(-\alpha)$. Hence, the previous equation can be written as

$$2\,[r^2{}_i(\alpha) + r^2{}_i(\beta) + r^2{}_i(\gamma)] = 1 \tag{3.90}$$

For $i = 1$, the previous equation reduces to

$$2\,[r^2{}_1(\alpha) + r^2{}_1(\beta) + r^2{}_1(\gamma)] = 1 \tag{3.91}$$

Substituting the values of first components of the roots $\boldsymbol{\alpha}$, $\boldsymbol{\beta}$, and $\boldsymbol{\gamma}$ from Equation (3.89) in Equation (3.91), we get

$$2(4b_1{}^2 + b^2{}_1 + b_1{}^2) = 1$$

or

$$b^2{}_1 = 1/12$$

or

$$b_1 = \pm\frac{1}{2\sqrt{3}} \tag{3.92}$$

Choosing the positive sign, we have

$$b_1 = \frac{1}{2\sqrt{3}}$$

Hence, three nonzero distinct roots of SU(3) are

$$\boldsymbol{\alpha} = \left(\frac{1}{\sqrt{3}}, 0\right)$$
$$\boldsymbol{\beta} = \left(\frac{1}{2\sqrt{3}}, \frac{1}{2}\right) \tag{3.93}$$
$$\boldsymbol{\gamma} = \left(\frac{1}{2\sqrt{3}}, -\frac{1}{2}\right)$$

The other three nonzero roots are $-\boldsymbol{\alpha}$, $-\boldsymbol{\beta}$, and $-\boldsymbol{\gamma}$.

PROBLEM 3.27
Comment on the statement that the choice of sign in the values of b_1 and b_2 is immaterial.

As already pointed out, there are two zero roots, each being represented by (0, 0). The six nonzero roots of SU(3) are shown in the root diagram (Figure 3.4) in which the components of a vector **r**, viz., r_1 and r_2, are taken as rectangular axes and the root vectors are drawn from a common origin. In the same diagram, the two zero vectors are shown by circles about the common origin.

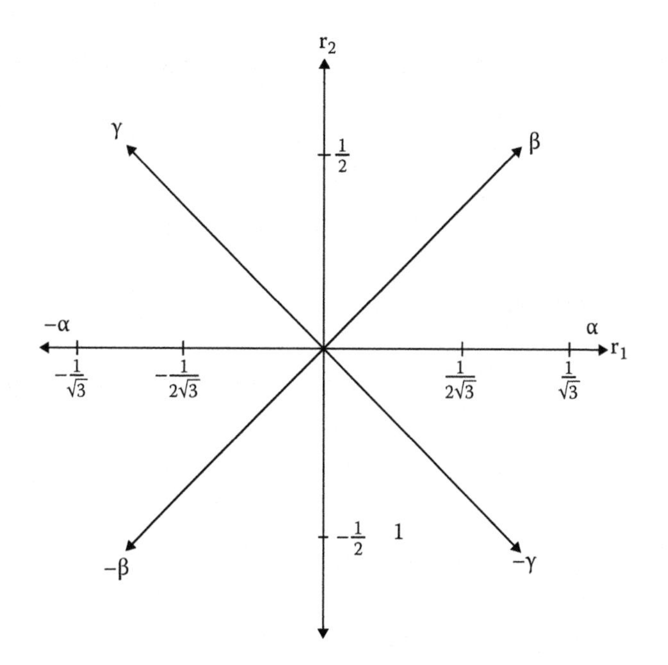

FIGURE 3.4
Root diagram for SU(3).

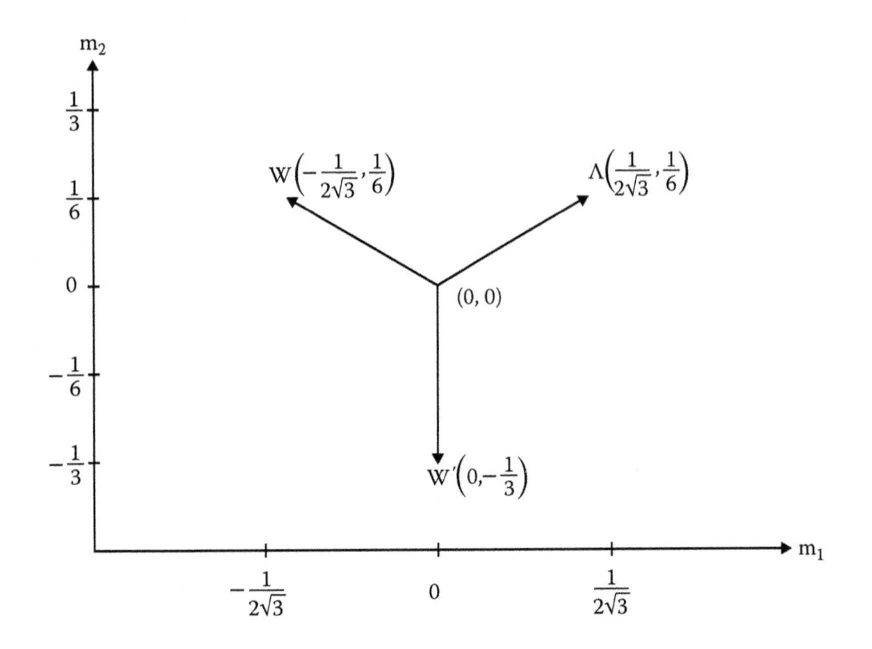

FIGURE 3.5
Weight diagram for $D^3(1, 0) = \{3\}$.

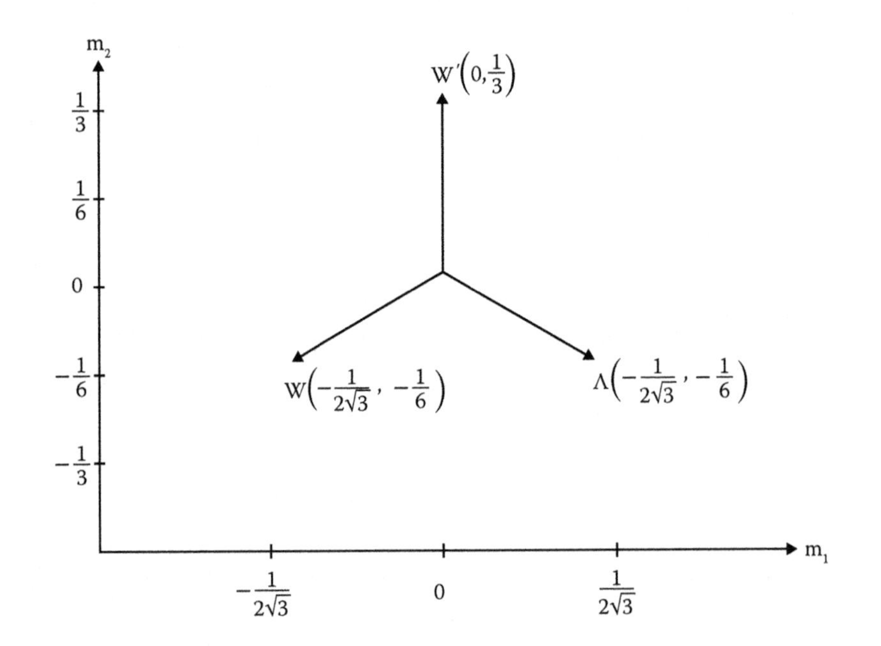

FIGURE 3.6
Weight diagram for $D^3(0, 1) = \{3^*\}$.

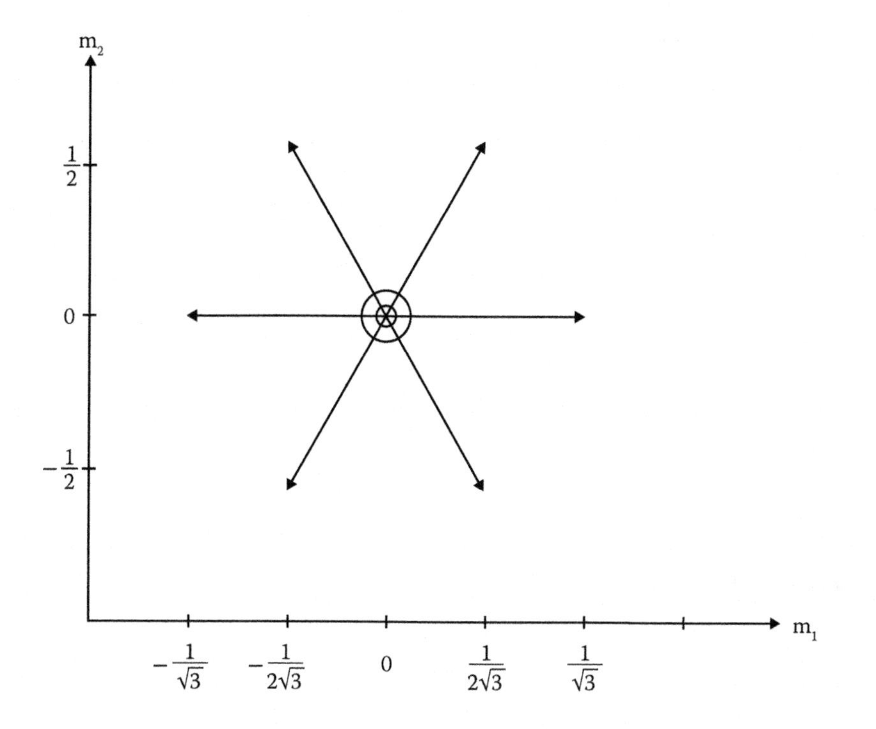

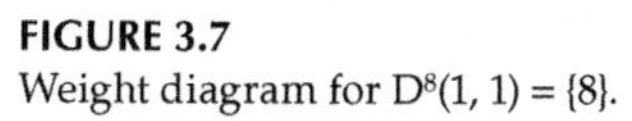
FIGURE 3.7
Weight diagram for $D^8(1, 1) = \{8\}$.

PROBLEM 3.28

Show that for $\varphi_{\alpha\beta} = 30°$, a Lie group of rank 2 has the following six distinct roots:

$$\boldsymbol{\alpha} = (1,\ 0)$$

$$\boldsymbol{\beta} = \left(\frac{3}{2}, \frac{\sqrt{3}}{2}\right)$$

$$\boldsymbol{\gamma} = \left(\frac{1}{2}, \frac{\sqrt{3}}{2}\right)$$

$$\boldsymbol{\delta} = \left(\frac{1}{2}, -\frac{\sqrt{3}}{2}\right)$$

$$\boldsymbol{\varepsilon} = (0, -\sqrt{3})$$

$$\boldsymbol{\rho} = \left(\frac{3}{2}, -\frac{\sqrt{3}}{2}\right)$$

3.20 Numerical Values of the Structure Constants of SU(3)

We will now determine the values of nonvanishing structure constants of SU(3). By inspection of the root diagram, we notice that while $\boldsymbol{\beta} + \boldsymbol{\gamma}$ is a root, the vector $\boldsymbol{\beta} + 2\boldsymbol{\gamma}$ is not a root. Therefore, $N_{\beta\gamma}$ exists and $j = 1$. Moreover, from the diagram, $\boldsymbol{\beta} - \boldsymbol{\gamma}$ is not a root, so that $k = 0$. Therefore, Equation (3.77a) gives

$$N_{\beta\gamma} = \pm\sqrt{\frac{1}{2}\, j\, (k+1)\boldsymbol{\gamma}.\boldsymbol{\gamma}} = \pm\frac{1}{\sqrt{6}} \tag{3.94}$$

We choose $N_{\beta\gamma} = +1/\sqrt{6}$. The values of other structure constants can be determined from symmetry relations.

3.21 Weights of a Representation

The concept of weights plays a significant role in the application of group theory to high energy physics. The weights are defined for every finite-dimensional representation of a group and are used to distinguish various such representations of the group.

Consider a Lie group G of order r and rank ℓ. Let $H_1, H_2, \ldots, H_\ell$ be the mutually commuting set of generators and $E_1, E_2, \ldots, E_{r-\ell}$ be the remaining set of generators of this group. We suppose that the group has been parameterized in such a manner that the generators are Hermitian. Since the Hermitian operators H_i, $i = 1, 2, \ldots, \ell$, mutually commute, there exists a complete set of simultaneous eigenkets of these operators. But the number of members of the complete set is determined only by the dimension of the space in which a representation of the group is considered. Let $|m>$ be *a* simultaneous eigenket of the operators H_i corresponding to the eigenvalues m_i in a vector space of dimension d. Then

$$H_i|m> = m_i|m>, \quad i = 1, 2, \ldots, \ell \tag{3.95}$$

That is,

$$H_1|m> = m_1|m>$$
$$H_2|m> = m_2|m>$$
$$\cdots\cdots\cdots\cdots$$
$$H_\ell|m> = m_\ell|m>$$

The eigenvalues $m_1, m_2, \ldots, m_\ell$ of the Hermitian generators $H_1, H_2, \ldots, H_\ell$ corresponding to a simultaneous eigenket $|m>$ can be considered as the components of a vector in an ℓ-dimensional space. This vector is called a *weight vector* or sometimes simply a *weight* of the given representation of the group or the *weight of the eigenstate* $|m>$. The ℓ-dimensional space is called the *weight space*. We denote this weight by

$$\mathbf{m} = (m_1, m_2, \ldots, m_\ell)$$

Thus, we can write

$$\mathbf{H}\ |m> = \mathbf{m}\ |m>$$

where $\mathbf{H} = (H_1, H_2, \ldots, H_\ell)$, and $\mathbf{m} = (m_1, m_2, \ldots, m_\ell)$. Since the operators H_i are Hermitian, the eigenvalues m_i are real. Hence, a weight of a representation is always real. Notice that the number of components of a weight is the same as the rank of a group.

Let us next see how many weights a representation of the group G in a space of dimension d can have. The maximum number of linearly independent ket vectors that form a complete set in this space is also d. Each one of these ket vectors will be a simultaneous eigenket of all the operators H_i and therefore give rise to a weight. Hence, the maximum number of distinct weights in this representation is d. However, as all the weights of a given representation are not necessarily distinct, in general the number of distinct weights will be less than d.

We will now give some definitions and prove or only state certain theorems that will be used later on.

Theorem 3.4

If $\mathbf{m}$ is a weight of a given representation of a semisimple Lie group and $\mathbf{r}(\alpha)$ is a root of the group, then $\mathbf{m} + \mathbf{r}(\alpha)$ is also a weight of that representation.

PROOF

Since $\mathbf{m}$ is a weight of a given representation of a group, we must have

$$H_i|m> = m_i|m> \tag{3.95'}$$

where $|m>$ is a simultaneous eigenvector of H_i.

We know that

$$[H_i, E_\alpha] = r_i(\alpha)\ E_\alpha$$

Operating this on a simultaneous eigenvector $|m>$ of all H_i, we have

$$[H_i, E_\alpha]|m> = r_i(\alpha)\ E_\alpha|m>$$

or

$$H_i E_\alpha |m> - E_\alpha H_i |m> = r_i(\alpha) E_\alpha |m>$$

or

$$H_i E_\alpha |m> - E_\alpha m_i |m> = r_i(\alpha) E_\alpha |\ m>$$

or

$$H_i(E_\alpha |m>) = [m_i + r_i(\alpha)](E_\alpha |m>) \tag{3.96}$$

where in the last but one step we have made use of the fact that m_i is the eigenvalue of H_i corresponding to the simultaneous eigenvector $|m>$. Equation (3.96) shows that $m_i + r_i(\alpha)$ is also an eigenvalue of H_i but corresponds to a different eigenvector $(E_\alpha |m>)$. Thus, the eigenvalues of the set $\mathbf{H} = (H_1, H_2, \ldots, H_\ell)$ form the weight $\mathbf{m} + \mathbf{r}(\alpha) = (m_1 + r_1(\alpha), m_2 + r_2(\alpha), \ldots, m_\ell + r_\ell(\alpha))$ associated with the eigenvector $(E_\alpha |m>)$. Hence, $\mathbf{m} + \mathbf{r}(\alpha)$ is also a weight of the same representation.

This theorem shows that the generator E_α serves as a "raising" operator; it "raises" the eigenvalue of H_i from m_i to $m_i + r_i(\alpha)$. Similarly, it can be shown that the generator $E_{-\alpha}$ serves as a "lowering" operator.

We now give two more definitions. A weight is said to be *simple* if it belongs to only one eigenstate $|m>$ of a representation. The number of linearly independent eigenvectors corresponding to the same weight of a representation is called the *multiplicity of the weight.* ■

Theorem 3.4

The eigenvectors belonging to *different weights* of the set $\mathbf{H} = (H_1, H_2, \ldots, H_\ell)$ of commuting generators are orthogonal to each other.

PROOF

Let $\mathbf{m}$ and $\mathbf{m}'$ be two distinct weights of $\mathbf{H}$ corresponding to the simultaneous eigenkets $|m>$ and $|m'>$. Then, we may write

$$\mathbf{H}|m> = \mathbf{m}|m> \tag{3.97}$$

and

$$\mathbf{H}|m'> = \mathbf{m}'|m'> \tag{3.98}$$

Taking the Hermitian conjugate of Equation (3.98) and noting that the eigenvalues of the Hermitian operator **H** are always real, we get

$$<m'|\mathbf{H}^{\dagger} = <m'|\mathbf{m}'^{*}$$

or

$$<m'|\mathbf{H} = \mathbf{m}'<m'| \tag{3.99}$$

The scalar product $<m'|\mathbf{H}|m>$ is obtained by multiplying Equation (3.97) from the left by $<m'|$ and Equation (3.99) from the right by $|m>$:

$$<m'|\mathbf{H}|m> = \mathbf{m}\ <m'|m>$$

and

$$<m'|\mathbf{H}|m> = \mathbf{m}'<m'|m>$$

Comparing these two equations, we have

$$(\mathbf{m} - \mathbf{m}')\ <m'|m> = 0$$

Since the two weights are distinct, $\mathbf{m} \neq \mathbf{m}'$, this equation yields

$$<m'|m> = 0$$

That is, the simultaneous eigenkets $|m>$ and $|m'>$ corresponding to different weights are orthogonal to each other. This proves thetheorem. ■

Theorem 3.6

If **m** is a weight and **α** is a root, then $g(m/\alpha) = 2(\mathbf{m} \cdot \boldsymbol{\alpha})/(\boldsymbol{\alpha} \cdot \boldsymbol{\alpha})$ is an integer and $\mathbf{m} - g(m/\alpha)\,\boldsymbol{\alpha}$ is also a weight with the same multiplicity as **m**.

PROOF

See Appendix D.

The weights related by this theorem are called *equivalent weights.* We now give some more definitions about the weights of a representation. A weight $\mathbf{m} = (m_1, m_2, \ldots, m_\ell)$ is said to be *positive* if its first nonzero component is positive. A weight $\mathbf{m} = (m_1, m_2, \ldots, m_\ell)$ is said to be higher than the weight $\mathbf{m}' = (m'_1, m_2', \ldots, m'_\ell)$ if the first nonzero component of $\mathbf{m} - \mathbf{m}'$ is positive. The highest member of a set of equivalent weights is said to be a *dominant weight.*

The *highest weight* of a representation is higher than all other weights of that representation. ■

> **REMARK**
>
> For SU(n), the following convention is also adopted. The weight $\mathbf{m}$ is higher than the weight $\mathbf{m}'$ if the last component of $\mathbf{m} - \mathbf{m}'$ is positive. If it is zero, then the next to the last component is considered. And so on.

We now state three theorems but omit their proofs.

Theorem 3.6

The highest weight of a representation is simple if and only if the representation is irreducible. ■

Theorem 3.7

Two irreducible representations (IRs) are equivalent if and only if they have the same highest weight. ■

Thus, an irreducible representation is uniquely characterized by its highest weight. That is, if $\mathbf{\Lambda}$ is the highest weight of an IR, no other IR can have $\mathbf{\Lambda}$ as its highest weight. If the highest weight is known, all other weights can be obtained by successive applications of the lowering operators $E_{-\alpha}$.

Theorem 3.8

A necessary and sufficient condition for a vector $\mathbf{\Lambda}$ to be the highest weight of some irreducible representation is that for a simple root $\mathbf{s}$,

$$g\left(\frac{\Lambda}{s}\right) = 2\frac{\mathbf{\Lambda} \cdot \mathbf{s}}{\mathbf{s} \cdot \mathbf{s}}$$

is a nonnegative integer. ■

The condition is necessary. It means that if Λ is the highest weight of some IR and **s** is a simple root, then

$$g\left(\frac{\Lambda}{s}\right) = 2\frac{\Lambda \cdot \mathbf{s}}{\mathbf{s} \cdot \mathbf{s}}$$

is a nonnegative integer.

Since this condition is also sufficient, it means that if **s** is a simple root and Λ is a vector such that

$$2\frac{\Lambda \cdot \mathbf{s}}{\mathbf{s} \cdot \mathbf{s}}$$

a nonnegative integer, then there must exist an irreducible representation with $\boldsymbol{\Lambda}$ as its highest weight.

This theorem helps us to calculate the highest weight of any irreducible representation.

3.22 Computation of the Highest Weight of Any Irreducible Representation of SU(3)

Let us calculate the highest weight $\boldsymbol{\Lambda} = (\Lambda_1, \Lambda_2)$ of any irreducible representation of SU(3). The three positive roots of SU(3) are given by Equations (3.95). Since SU(3) is a group of rank $\ell = 2$, it has two simple roots:

$$\boldsymbol{\beta} = \left(\frac{1}{2\sqrt{3}}, \frac{1}{2}\right)$$

and

$$\boldsymbol{\gamma} = \left(\frac{1}{2\sqrt{3}}, -\frac{1}{2}\right)$$

Notice that $\boldsymbol{\alpha}$ is not a simple root as it can be expressed as a sum of two positive roots $\boldsymbol{\beta}$ and $\boldsymbol{\gamma}$. Now, for the simple root $\boldsymbol{\beta}$, we have

$$2\left(\frac{\boldsymbol{\Lambda} \cdot \boldsymbol{\beta}}{\boldsymbol{\beta} \cdot \boldsymbol{\beta}}\right) = 2\left(\frac{\Lambda_1}{2\sqrt{3}}, \frac{\Lambda_2}{2}\right) \Big/ \frac{1}{3} = \sqrt{3}\Lambda_1 + 3\Lambda_2$$

For an irreducible representation, this must be a nonnegative integer, that is,

$$\sqrt{3}\Lambda_1 + 3\Lambda_2 = a$$

where a is a nonnegative integer. Similarly, for the simple root $\boldsymbol{\gamma}$, we have

$$2\frac{\boldsymbol{\Lambda}\cdot\boldsymbol{\gamma}}{\boldsymbol{\gamma}\cdot\boldsymbol{\gamma}} = \sqrt{3}\Lambda_1 - 3\Lambda_2 = b$$

where b is also a nonnegative integer.

Solving these two equations for Λ_1 and Λ_2, we get

$$\Lambda_1 = \frac{1}{2\sqrt{3}}(a+b), \quad \Lambda_2 = \frac{1}{6}(a-b)$$

Thus,

$$\begin{aligned}\boldsymbol{\Lambda} &= (\Lambda_1,\ \Lambda_2)\\ &= \left(\frac{1}{2\sqrt{3}}(a+b),\ \frac{1}{6}(a-b)\right)\end{aligned}$$

where a and b are nonnegative integers. Hence, $\boldsymbol{\Lambda}$, the highest weight of any irreducible representation of SU(3), has two components and is given by the previous equation. It is expressed in terms of two nonnegative integers a and b. The values assigned to these integers determine the highest weight of some irreducible representations of SU(3). The dimension of any IR of SU(3) in terms of a and b can be obtained by the following theorem due to Weyl.

Weyl's Theorem 3.9

If $\boldsymbol{\Lambda}$ is the highest weight of an irreducible representation of a semisimple Lie group, then the dimension d of this representation is given by

$$d = \prod_{\mathbf{R}_+}[1 + (\boldsymbol{\Lambda}\cdot\mathbf{R}_+/(\mathbf{g}\cdot\mathbf{R}_+)$$

where $\mathbf{R}_+$ is a positive root of the semisimple Lie group and

$$\mathbf{g} = \frac{1}{2}\sum_{\mathbf{R}_+}\mathbf{R}_+$$

■

Example 3.5

Let us derive the formula for the dimensionality of any irreducible representation of SU(3). The positive roots of SU(3) are

$$\boldsymbol{\alpha} = \left(\frac{1}{\sqrt{3}},\ 0\right)$$

$$\boldsymbol{\beta} = \left(\frac{1}{2\sqrt{3}},\ \frac{1}{2}\right)$$

$$\boldsymbol{\gamma} = \left(\frac{1}{2\sqrt{3}},\ \frac{1}{2}\right)$$

Therefore, $\sum_{\mathbf{R}_+} \mathbf{R}_+$, the vector sum of three positive roots of the group is given by

$$\sum_{\mathbf{R}_+} \mathbf{R}_+ = (\boldsymbol{\alpha} + \boldsymbol{\beta} + \boldsymbol{\gamma})$$

$$= \left(\frac{2}{\sqrt{3}},\ 0\right)$$

$$= 2\boldsymbol{\alpha}$$

This shows that

$$\mathbf{g} = \frac{1}{2}\sum_{\mathbf{R}_+} \mathbf{R}_+$$

$$= \left(\frac{1}{\sqrt{3}}, 0\right)$$

$$= \boldsymbol{\alpha}$$

This yields

$$\mathbf{g} \cdot \boldsymbol{\alpha} = \frac{1}{3}$$

$$\mathbf{g} \cdot \boldsymbol{\beta} = \frac{1}{6}$$

$$\mathbf{g} \cdot \boldsymbol{\gamma} = \frac{1}{6}$$

Moreover, it has already been shown that

$$\boldsymbol{\Lambda} = \left(\frac{1}{2\sqrt{3}}(a+b),\ \frac{1}{6}(a-b)\right)$$

Consequently,

$$\mathbf{\Lambda} \cdot \boldsymbol{\alpha} = \frac{1}{\sqrt{3}}\Lambda_1 = \frac{1}{6}(a+b)$$

$$\mathbf{\Lambda} \cdot \boldsymbol{\beta} = \frac{1}{2\sqrt{3}}\Lambda_1 + \frac{1}{2}\Lambda_2 = \frac{1}{6}a$$

$$\mathbf{\Lambda} \cdot \boldsymbol{\gamma} = \frac{1}{6}b.$$

Now, by virtue of Weyl's theorem 3.10, the dimension d of any irreducible representation of SU(3) is given by

$$d = \left(1 + \frac{\mathbf{\Lambda} \cdot \boldsymbol{\alpha}}{\mathbf{g} \cdot \boldsymbol{\alpha}}\right)\left(1 + \frac{\mathbf{\Lambda} \cdot \boldsymbol{\beta}}{\mathbf{g} \cdot \boldsymbol{\beta}}\right)\left(1 + \frac{\mathbf{\Lambda} \cdot \boldsymbol{\gamma}}{\mathbf{g} \cdot \boldsymbol{\gamma}}\right)$$

Substituting the expressions for various scalar products and simplifying, we get

$$d = \left(1 + \frac{a+b}{2}\right)(1+a)(1+b)$$

where a and b are nonnegative integers identifying any irreducible representation. We can, therefore, label an irreducible representation by (a, b) or $D^d(a, b)$ or $D^{(d)}(a, b)$. Thus the representation corresponding to a = 1, b = 0 is denoted by $D^3(1, 0)$, whereas $D^3(0, 1)$ denotes the representation characterized by a = 0, b = 1. For the sake of brevity, it is usual to write [3], **3**, or 3 for $D^3(1, 0)$. When two or more irreducible representations have the same dimensions, we can distinguish them by, for example, stars, primes, or bars. For example, since $D^3(1, 0)$ and $D^3(0, 1)$ have the same dimension, the latter is denoted by 3* or $\bar{3}$. This representation is known as the conjugate of 3. In general, a bar or an asterisk on a number is used to denote the conjugate of the corresponding representation.

In literature, the nonnegative integers are denoted more frequently by a_1 and a_2. Then the dimension of an IR of SU(3) may be written as

$$d = \left(1 + \frac{a_1 + a_2}{2}\right)(1+a_1)(1+a_2)$$

Sometimes, it is required to express the dimension of any IR of SU(3) by two nonnegative integers λ_1 and λ_2 such that $\lambda_1 \geq \lambda_2$. Then, defining

$$\lambda_1 = a_1 + a_2 \quad \text{and} \quad \lambda_2 = a_2$$

so that λ_1 is always either equal to or greater than λ_2, we may write

$$d = \left(1 + \frac{\lambda_1}{2}\right)(1 + \lambda_1 - \lambda_2)(1 + \lambda_2)$$

It is significant that some Lie groups can have irreducible representations of finite dimensions.

PROBLEM 3.29

Show that the dimension of any irreducible representation of

(a) SU(2) is given by

$$d = 1 + a$$

(b) C_2 is given by

$$d = (1+a)(1+b)\left(1 + \frac{a+b}{2}\right)\left(1 + \frac{a+2b}{3}\right)$$

(c) G_2 is given by

$$d = (1+a)(1+b)\left(1+\frac{a+b}{2}\right)\left(1+\frac{a+2b}{3}\right)$$
$$\times\left(1+\frac{a+3b}{4}\right)\left(1+\frac{2a+3b}{5}\right)$$

3.23 Dimension of any Irreducible Representation of SU(N)

Sometimes, the set of nonnegative integers $\lambda_1, \lambda_2, \ldots, \lambda_{N-1}$ where $\lambda_1 \geq \lambda_2 \geq \cdots \geq \lambda_{N-1}$, is defined in terms of the set of nonnegative integers $a_1, a_2, \ldots, a_{N-1}$ by the following equations:

$$\lambda_1 = a_1 + a_2 + a_3 + \cdots + a_{N-1}$$
$$\lambda_2 = a_1 + a_2 + a_3 + \cdots + a_{N-2}$$
$$\cdots\cdots\cdots\cdots\cdots\cdots$$
$$\lambda_{N-2} = a_1 + a_2$$
$$\lambda_{N-1} = a_1$$

Then an IR of SU(N) is specified by $(\lambda_1, \lambda_2, \ldots, \lambda_{N-1})$ and its dimension d is given by

$$
\begin{gathered}
d(\lambda_1, \lambda_2, \ldots, \lambda_{N-1}) \\
= (\lambda_1 - \lambda_2 + 1)(\lambda_1 - \lambda_3 + 2) \ldots \\
\{\lambda_1 - \lambda_{N-1} + (N-1)\}(\lambda_1 + N - 1) \times \\
(\lambda_2 - \lambda_3 + 1)(\lambda_2 - \lambda_4 + 2) \ldots \{\lambda_2 - \lambda_{N+1} + (N-3)\} \\
\times (\lambda_2 + N - 2) \\
\cdots\cdots\cdots\cdots\cdots\cdots\cdots\cdots\cdots\cdots \\
\times (\lambda_{N-2} - \lambda_{N-1} + 1)(\lambda_{N-2} + 2) \\
\times(\lambda_{N-1} + 1)/[1!2! \ldots (N-1)!]
\end{gathered}
$$

Therefore, the formula for determining the dimension of any IR of SU(2) is

$$d(\lambda_1) = \lambda_1$$

For any IR of SU(3), the corresponding formula for the dimension is

$$d(\lambda_1, \lambda_2) = \frac{1}{2} - (\lambda_1 - \lambda_2 + 1)(\lambda_1 + 2)(\lambda_2 + 1)$$

and

for the IR of SU(4), we have

$$
\begin{aligned}
d(\lambda_1, \lambda_2, \lambda_3) = \frac{1}{2!}\,\frac{1}{3!}\,&(\lambda_3 + 1)(\lambda_2 - \lambda_3 + 1)(\lambda_2 + 2) \\
&\times (\lambda_1 - \lambda_2 + 1)(\lambda_1 - \lambda_3 + 2)(\lambda_1 + 3)
\end{aligned}
$$

For SU(3), in terms of a_1 and a_2, the formula for the dimension of any IR may be written as

$$
\begin{aligned}
d(a_1,\ a_2) &= \frac{1}{2}(a_1 + a_2 - a_1 + 1)(a_1 + a_2 + 2)(a_1 + 1) \\
&= \left(1 + \frac{a_1 + a_2}{2}\right)(1 + a_1)(1 + a_2)
\end{aligned}
$$

3.24 Computation of the Weights of Any Irreducible Representation of SU(3)

We have already shown that the highest weight of any irreducible representation of SU(3) is given by

$$\mathbf{\Lambda} = \left(\frac{1}{2\sqrt{3}}(a+b),\ \ \frac{1}{6}(a-b) \right)$$

and characterizes the irreducible representation. However, for physical applications, it is necessary to determine all the weights of an IR. Therefore, by making use of this result, we shall calculate all the weights of $D^3(1, 0)$ of SU(3). The values of a and b for this representation are 1 and 0, so that the highest weight of this representation is

$$\mathbf{\Lambda} = \left(\frac{1}{2\sqrt{3}},\ \frac{1}{6} \right)$$

Now, since $\mathbf{\Lambda}$ is a weight of the IR $D^3(1, 0)$ and $\boldsymbol{\alpha}$ is a root of SU(3), then $\mathbf{\Lambda} - g(\Lambda/\alpha)\boldsymbol{\alpha}$, where $g(\Lambda/\alpha) = 2\,(\mathbf{\Lambda} \cdot \boldsymbol{\alpha})/(\boldsymbol{\alpha} \cdot \boldsymbol{\alpha})$ must also be a weight of $D^3(1, 0)$. Since $g(\Lambda/\alpha) = 1$, this weight is given by

$$\mathbf{\Lambda} - g(\Lambda/\alpha)\boldsymbol{\alpha} = \left(\frac{1}{2\sqrt{3}},\ \frac{1}{6} \right) - \left(\frac{1}{\sqrt{3}},\ 0 \right) = \left(-\frac{1}{2\sqrt{3}},\ \frac{1}{6} \right) = \mathbf{\Lambda},\ \ \text{say.}$$

Similarly,

$$\mathbf{\Lambda} - g(\Lambda/\beta)\boldsymbol{\beta} = \mathbf{\Lambda} - \boldsymbol{\beta} = \left(0,\ -\frac{1}{3} \right) = \mathbf{W}',\ \ \text{say,}$$

the third weight of $D^3(1, 0)$. The irreducible representation $D^3(1, 0)$, being of dimension 3, can have at the most three weights. All the weights of this representation have therefore been determined. These are

$$\mathbf{\Lambda} = \left(\frac{1}{2\sqrt{3}},\ \frac{1}{6} \right)$$

$$\mathbf{W} = \left(-\frac{1}{2\sqrt{3}},\ \frac{1}{6} \right)$$

$$\mathbf{W}' = \left(0,\ -\frac{1}{3} \right)$$

It may be noticed that $\mathbf{\Lambda} - g(\Lambda/\gamma)\boldsymbol{\gamma}$ does not give a new weight as $g(\Lambda/\gamma) = 0$.

REMARK

These values of the weights are different from those obtained by some authors as their root normalization is different.

PROBLEM 3.30

Show that the weights of the IR representation 3* of SU(3) are

$$\mathbf{\Lambda} = \left(\frac{1}{2\sqrt{3}}, \ -\frac{1}{6} \right)$$

$$\mathbf{W} = \left(-\frac{1}{2\sqrt{3}}, \ -\frac{1}{6} \right)$$

$$\mathbf{W}' = \left(0, \ \frac{1}{3} \right)$$

It is important to point out that the technique used here for obtaining all the weights of $D^3(1, 0)$ has been successful only because all the weights of $D^3(1, 0)$ are equivalent. If all the weights of a representation are not equivalent, this method fails. For instance, the groups C_2 and G_2, which are also simple groups of rank $\ell = 2$, do not possess this property. Therefore, we cannot obtain all the weights of their irreducible representations by the aforementioned technique. A method that is applicable in all cases is given as follows.

Suppose that we want to calculate the weights of the irreducible representation $D^d(a, b)$ of SU(3). Let $\mathbf{m}$ be any one of its weights. Since the rank ℓ of the group is 2, it will have two simple roots. Let $\mathbf{s}$ be one of the simple roots of SU(3). Then, write down the series

$$\mathbf{m}, \mathbf{m} + \mathbf{s}, \mathbf{m} + 2\mathbf{s}, \ldots$$

where the first term is the given weight and successive terms of the series are obtained by adding the simple root $\mathbf{s}$. Retain only those terms in the series which are weights of this IR, writing the last term as $\mathbf{m} + q\,\mathbf{s}$, where q is a nonnegative integer, such that $\mathbf{m} + q\,\mathbf{s}$ is a weight but $\mathbf{m} + (q + 1)\mathbf{s}$ is not a weight. That is, there are $q + 1$ weights and consequently $q + 1$ terms in the series. Next compute the number $p \equiv [q + g(m/s)]$, where, as already defined,

$$g(m/s) = 2\,\frac{\mathbf{m} \cdot \mathbf{s}}{\mathbf{s} \cdot \mathbf{s}}$$

If $p > 0$, then it can be shown that $\mathbf{m} - \mathbf{s}$, where $\mathbf{m}$ is the first weight in the series and $\mathbf{s}$ is a simple root, is another weight of the irreducible representation; otherwise it is not.

This process can be repeated with all other simple roots and also by selecting new weights calculated by this technique till all the weights of the irreducible representation are obtained.

We will illustrate this method by finding all the weights of $D^8(1,1)$ of SU(3).

3.25 Weights of Irreducible Representation $D^8(1,1)$ of SU(3)

Let Λ be the highest weight of irreducible representation $D^8(1,1)$ of SU(3). By virtue of the formula

$$\boldsymbol{\Lambda} = \left(\frac{1}{2\sqrt{3}}(a+b),\ \frac{1}{6}(a-b) \right)$$

this is given by

$$\boldsymbol{\Lambda} = \left(\frac{1}{\sqrt{3}},\ 0 \right) = \boldsymbol{\alpha}$$

as for the IR 8, $a = b = 1$. The group SU(3) has two simple roots $\boldsymbol{\beta}$ and $\boldsymbol{\gamma}$, which are given by

$$\boldsymbol{\beta} = \left(\frac{1}{2\sqrt{3}},\ \frac{1}{2} \right)$$

$$\boldsymbol{\gamma} = \left(\frac{1}{2\sqrt{3}},\ -\frac{1}{2} \right)$$

Let us write down the series with $\boldsymbol{\Lambda}$ as its first weight and $\boldsymbol{\beta}$ as a simple root. Then, we have the series

$$\boldsymbol{\Lambda},\ \boldsymbol{\Lambda} + \boldsymbol{\beta},\ \boldsymbol{\Lambda} + 2\boldsymbol{\beta},\ \ldots$$

Since $\boldsymbol{\Lambda}$ is the highest weight and $\boldsymbol{\beta} = \left(1/2\sqrt{3},\ 1/2\right)$ is a positive root (a simple root, by definition, is always positive), it follows that the vector $\boldsymbol{\Lambda} + \boldsymbol{\beta}$ is higher than the highest weight $\boldsymbol{\Lambda}$ and therefore cannot be a weight of $D^8(1,1)$, as no weight can be higher than the highest weight. Consequently, the series will contain only one member $\boldsymbol{\Lambda}$ that is a weight; the next term $\boldsymbol{\Lambda} + \boldsymbol{\beta}$ is not a weight. Let us now find out whether p as previously defined is greater than zero. If it is, then $\mathbf{\Lambda} - \boldsymbol{\beta}$ is also a weight. To determine p, we notice that since there is only one term in the series, we have $q + 1 = 1$ or $q = 0$. Moreover, $g(\Lambda/\beta) = 2\,\mathbf{\Lambda} \cdot \boldsymbol{\beta}/(\boldsymbol{\beta} \cdot \boldsymbol{\beta}) = 1$, so that $p = [q + g(\Lambda/\beta)] = 1 > 0$. Hence, $\mathbf{\Lambda} - \boldsymbol{\beta} = \boldsymbol{\alpha} - \boldsymbol{\beta} = \boldsymbol{\gamma}$ is also a weight of $D^8(1,1)$.

Similarly, writing down the series with Λ as the first weight and γ as a simple root, it can be shown that $\mathbf{\Lambda} - \boldsymbol{\gamma} = \boldsymbol{\alpha} - \boldsymbol{\gamma} = \boldsymbol{\beta}$ is also a weight of $D^8(1,1)$.

$(\mathbf{\Lambda} - \boldsymbol{\beta}) = \boldsymbol{\gamma}$ has been found already as a weight of $D^8(1,1)$. Let us now write down a series with the weight $\mathbf{\Lambda} - \boldsymbol{\beta}$ as the first term and $\boldsymbol{\beta}$ as a simple root. Then, we may write the series as

$$\mathbf{\Lambda} - \boldsymbol{\beta}, \mathbf{\Lambda}, \mathbf{\Lambda} + \boldsymbol{\beta}, \mathbf{\Lambda} + 2\boldsymbol{\beta}, \ldots$$

Since $\mathbf{\Lambda} + \boldsymbol{\beta}$, being higher than the highest weight Λ, cannot be a weight of $D^8(1,1)$, this series must terminate after two terms. That is, $q + 1 = 2$, which yields $q = 1$. The number $g[(\Lambda - \beta)/\beta]$ is given by

$$g\left(\frac{(\Lambda-\beta)}{\beta}\right) = 2\frac{(\mathbf{\Lambda} - \boldsymbol{\beta}).\,\beta}{\boldsymbol{\beta} \cdot \boldsymbol{\beta}} = 2\frac{\mathbf{\Lambda} \cdot \boldsymbol{\beta}}{\boldsymbol{\beta} \cdot \boldsymbol{\beta}} - 2 = 1 - 2 = -1$$

so that $p = [q + g((\Lambda - \beta)/\beta)] = 1 - 1 = 0 > 0$ and thus $(\mathbf{\Lambda} - \boldsymbol{\beta}) - \boldsymbol{\beta} = \mathbf{\Lambda} - 2\boldsymbol{\beta}$ is not a weight of $D^8(1,1)$. Similarly, it can be shown that $\mathbf{\Lambda} - 2\,\boldsymbol{\gamma}$ is not a weight of $D^8(1,1)$.

Let us next start with the same weight $\mathbf{\Lambda} - \boldsymbol{\beta}$ but write the series with $\boldsymbol{\gamma}$ as a simple root. We then have the series as

$$\mathbf{\Lambda} - \boldsymbol{\beta}, \mathbf{\Lambda} - \boldsymbol{\beta} + \boldsymbol{\gamma}, \mathbf{\Lambda} - \boldsymbol{\beta} + 2\,\boldsymbol{\gamma}, \ldots$$

We know that $\mathbf{\Lambda} - \boldsymbol{\beta}$ is a weight, but we do not know whether $\mathbf{\Lambda} - \boldsymbol{\beta} + \boldsymbol{\gamma}$ is a weight. If it is not a weight, the series must terminate after the first term, and we must have $q = 0$. Otherwise, we have to consider the next term and so on so that $q > 0$. That is in all cases, $q \geq 0$. The number $g((\Lambda - \beta)/\gamma)$ is given by

$$g\left(\frac{(\Lambda-\beta)}{\gamma}\right) = 2\frac{(\mathbf{\Lambda} - \boldsymbol{\beta}) \cdot \gamma}{\boldsymbol{\gamma} \cdot \boldsymbol{\gamma}} = 2\frac{\boldsymbol{\gamma} \cdot \boldsymbol{\gamma}}{\boldsymbol{\gamma} \cdot \boldsymbol{\gamma}} = 2$$

Thus, $p = [q + g((\Lambda - \beta)/\gamma)] = (\geq 0 + 2) \geq 2 > 0$, which shows that $\mathbf{\Lambda} - \boldsymbol{\beta} - \boldsymbol{\gamma} = \boldsymbol{\gamma} - \boldsymbol{\gamma} = \mathbf{0}$ is a weight of $D^8(1,1)$. Similarly, we can show that $\mathbf{\Lambda} - \boldsymbol{\gamma} - \boldsymbol{\beta} = \boldsymbol{\beta} - \boldsymbol{\beta} = \mathbf{0}$ is also a weight of $D^8(1,1)$. Thus, there are two zero weights.

Let us next start the series with zero as its first weight and β as a simple root. Then, the series is

$$\mathbf{0}, \boldsymbol{\beta}, 2\boldsymbol{\beta}, 3\boldsymbol{\beta}, \ldots$$

We have already proved that $\boldsymbol{\beta}$ is also a weight of the IR 8. We do not know whether $2\boldsymbol{\beta}$ is a weight. Therefore, as before, $q \geq 1$. Moreover, $g(0/\beta) = 2\frac{\mathbf{0}\cdot\boldsymbol{\beta}}{\boldsymbol{\beta}\cdot\boldsymbol{\beta}} = 0$. Thus, $p = [q + g(0/\beta)] = (\geq 1 + 0) \geq 1 > 0$. Hence, $\mathbf{0} - \boldsymbol{\beta} \equiv -\boldsymbol{\beta}$ is also a weight of the IR 8. Similarly, starting with $\mathbf{0}$ as a weight and choosing γ as a simple root, we can show that $-\boldsymbol{\gamma}$ is also a weight.

Finally, starting with the weight $-\boldsymbol{\beta}$ and taking $\boldsymbol{\gamma}$ as a simple root, it can be shown that $-\boldsymbol{\beta} - \boldsymbol{\gamma} \equiv -\boldsymbol{\alpha}$ is also a weight.

Since, the IR is of dimension 8, the maximum number of its weights can be eight. All these weight vectors or weights have been determined. These are $\boldsymbol{\alpha}, \boldsymbol{\beta}, \boldsymbol{\gamma}, -\boldsymbol{\alpha}, -\boldsymbol{\beta}, -\boldsymbol{\gamma}, \mathbf{0}$ and $\mathbf{0}$. It may be noticed that the weights of the irreducible representation 8 of SU(3) are the same as the roots of the group SU(3).

It can be proved that the weights around the periphery of a weight diagram are simple; the others may be multiple. The maximum multiplicity v of any weight of a weight diagram is given by

$$v = \frac{1}{2}(a_1 + a_2) - \frac{1}{2}|a_1 - a_2| + 1$$

now state an important theorem that is due to Cartan.

Cartan's Theorem 3.10

For every simple group of rank ℓ, there are ℓ dominant weights, called *fundamental dominant weights*, $\mathbf{M}_i$, $i = 1, 2, \ldots, \ell$, such that any other dominant weight $\mathbf{M}$ is a linear combination of them:

$$\mathbf{M} = \sum_{i=1}^{\ell} a_i \mathbf{M}_i$$

where a_i, called the *Cartan parameters*, are nonnegative integers. Furthermore, there exist ℓ fundamental irreducible representations that have ℓ fundamental dominant weights as their highest weights. ■

As an illustration, let us consider the group SU(3). Its rank ℓ is two, and it has two fundamental dominant weights that, according to Theorem 3.10, are the highest weights of two irreducible representations (IRs).

Now the highest weight of any irreducible representation of SU(3) is given by

$$\boldsymbol{\Lambda} = \left(\frac{1}{2\sqrt{3}}(a + b), \ \frac{1}{6}(a - b) \right)$$

where a and b are nonnegative integers. This equation can be written as

$$\Lambda = \left(\frac{1}{2\sqrt{3}} a + \frac{1}{2\sqrt{3}} b,\ \frac{1}{6} a - \frac{1}{6} b \right)$$

$$= a\left(\frac{1}{2\sqrt{3}}, \frac{1}{6} \right) + b\left(\frac{1}{2\sqrt{3}}, -\frac{1}{6} \right) = a\mathbf{M}_1 + b\ \mathbf{M}_2$$

The vectors $\mathbf{M}_1$ and $\mathbf{M}_2$ are the highest weights of IRs, (1, 0) and (0, 1), respectively. The last relation shows that the highest weight of any IR of SU(3) can be expressed as a linear combination of $\mathbf{M}_1$ and $\mathbf{M}_2$ with nonnegative integral coefficients. Hence, $\mathbf{M}_1$ and $\mathbf{M}_2$ are two fundamental weights and (1, 0) and (0, 1) are two fundamental representations of SU(3).

The *fundamental representations* are the representations of the lowest dimensions (1 excluded). This fact may be used to determine the fundamental representations by using the dimensionality formula. For SU(3), the dimension of any IR is given by

$$d = (1 + a)(1 + b)\left(1 + \frac{a + b}{2} \right)$$

where a and b are nonnegative integers. For $a = 0 = b$, we have $d = 1$. This is excluded. From among the remaining representations, the two with the lowest dimension are (1, 0) and (0, 1), that is, 3 and 3*. These are the fundamental representations of SU(3).

The dimensionality d of any IR of C_2 is given by

$$d = (1 + a_1)(1 + a_2)\left(1 + \frac{a_1 + a_2}{2} \right)\left(1 + \frac{a_1 + 2a_2}{3} \right)$$

where a_1 and a_2 are nonnegative integers. The fundamental representations correspond to $a_1 = 1$, $a_2 = 0$ and $a_1 = 0$, $a_2 = 1$, that is, are $D^4(1,0)$ and $D^5(0,1)$.

PROBLEM 3.31
Show that $D^7(1, 0)$ and $D^{14}(0, 1)$ are fundamental representations of G_2.

3.26 Weight Diagrams

The graphical representation of weight vectors of any irreducible representation of a semi-simple Lie group of rank ℓ, in an ℓ-dimensional space is called a *weight diagram*. Weight diagrams for three irreducible representations 3, 3*, and 8 of SU(3) have been drawn on the next page. The components m_1 and m_2 of a weight have been taken along the two axes of a rectangular coordinate system in a two-dimensional space and the weights are plotted. The weight vectors are represented by lines drawn from the common origin to the weight points. The weight diagrams are of direct physical interest.

It is important to note that there is only one root diagram for every group, but each irreducible representation of the group has its own weight diagram.

Notice that the weight diagrams for two fundamental representations of SU(3) exhibit symmetry under rotation of 120° about the common origin. This occurs because of the symmetry of the root diagram and is therefore true for all representations.

REMARK

It can be proved that the weights around the periphery of a weight diagram are simple; the others may be multiple. The maximum multiplicity *m* of any weight is given by

$$m = \frac{1}{2}(a_1 + a_2) - |a_1 - a_2| + 1$$

3.27 Decomposition of a Product of Two Irreducible Representations

Chapter 2 showed that the direct product of two IRs of a group is itself a representation of the same group. This product representation is in general reducible and can be decomposed into the direct sum of IRs of the group. Here, we will not prove any theorem but will describe only two methods by which the decomposition of the *Kronecker* or *outer* or *direct product* of IRs of a unimodular unitary group, SU(n), can be achieved.

3.27.1 First Method

We will describe a method that is valid for the decomposition of the direct product of two IRs of SU(3) only. Suppose that $D^{(1)}$ and $D^{(2)}$ are two irreducible representations of SU(3). Let the pairs of nonnegative integers (a_1, a_2) and (a'_1, a'_2) characterize the representations $D^{(1)}$ and $D^{(2)}$, respectively. We will then determine the IRs of SU(3) contained in the direct product $D(a_1, a_2) \times D(a'_1, a'_2)$, also written as $(a_1, a_2) \times (a'_1, a'_2)$. The direct product of two IRs $D^{(1)}$ and $D^{(2)}$ can be decomposed into a direct sum of irreducible representations according to

$$(a_1, a_2) \times (a'_1, a'_2) = \sum_{i=0}^{\min(a_1, a'_2)} \sum_{j=0}^{\min(a'_1, a_2)} (a_1 - i, a_2 - j;\ a_1^* - j, a_2^* - j) \tag{3.100}$$

The upper limit for the first summation, namely, min (a_1, a'_2), means "a_1 or a'_2 whichever is smaller." This is similar for the second summation. We sum over i and j. Then each term on the right-hand side of relation (3.100) is decomposed in accordance with the following prescription:

$$(a_1, a_2; a'_1, a'_2) = (a_1 + a'_1, a_2 + a'_2)$$

$$+\sum_{i=1}^{\min(a_1, a'_2)} (a_1 + a'_1 - 2i,\ a_2 + a'_1 + i) + \sum_{j=1}^{\min(a'_2, a_2)} (a_1 + a'_1 - 2j,\ a_2 + a'_1 + j) \tag{3.101}$$

Terms on the right-hand side of this relation characterize the final irreducible representations contained in the product of $D(a_1, a_2)$ and $D(a'_1, a'_2)$.

To illustrate this technique, let us find the IRs contained in the Kronecker product of two IRs, [8] and [10] of SU(3). By virtue of the formula given earlier, these IRs are characterized by (1, 1) and (3,0), respectively. Therefore, relation (3.100) gives

$$(1, 1) \times (3, 0) = \sum_{i=0}^{0} \sum_{j=0}^{1} (1 - i,\ 1 - j;\ 3 - j,\ -i)$$

$$= \sum_{j=0}^{1} (1,\ 1 - j;\ 3 - j, 0) = (1,\ 1; 3,\ 0) + (1,\ 0; 2,\ 0) \tag{3.102}$$

By virtue of relation (3.101), we obtain

$$(1, 1; 3, 0) = (4, 1) + \sum_{j=1} (4 - 2j,\ 1 + j) = (4,\ 1) + (2, 2)$$

Similarly, we can show that

$$(1,0; 2,0) = (3,0) + (1, 1)$$

Hence, Equation (3.102) gives

$$(1, 1) \times (3, 0) = (4, 1) + (2, 2) + (3, 0) + (1, 1)$$

or

$$8 \times 10 = 35 + 27 + 10 + 8$$

PROBLEM 3.32
Show that

$$8 \times 8 = 1 + 8 + 8 + 10 + 10^* + 27$$

For evaluating a double product of SU(3), like (3 x 3) x 3, first of all the first product (3 x 3) is decomposed according to the previously outlined procedure, and the IRs contained in it are obtained. These IRs are multiplied in turn by the third member of the double product, and the aforementioned procedure for the reduction is repeated. Thus,

$$(3 \times 3) \times 3 = (3^* + 6) \times 3 = 3^* \times 3 + 6 \times 3 = 1 + 8 + 8 + 10$$

PROBLEM 3.33
Find the IRs contained in (3 x 3*) x 3* of SU(3).

3.27.2 Second Method

The first method is valid only for SU(3). We will now describe a procedure that can be used to find all the IRs contained in the Kronecker product of IRs of any *unimodular unitary group,* that is, of SU(n). This procedure involves the technique of Young's tableaux (YTs). We will, therefore, first explain the meaning of a YT. Consider a set of nonnegative integers $\lambda_1, \lambda_2, \ldots, \lambda_r$, such that $\lambda_1 \geq \lambda_2 \geq \cdots \geq \lambda_r$. Draw a diagram such that the first row contains λ_1 boxes, the second row contains λ_2 boxes, and so on, and the last row contains λ_r boxes, each row being left justified. Such a diagram is called YT. The YT for $\lambda_1 = 3$, $\lambda_2 = 1$ is

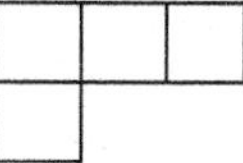

To find the IRs that occur in the outer product (also called Kronecker or direct product) of two IRs, we give the *prescription* with reference to the outer product of 8 x 8 of SU(3):

1. We note that the dimension of any IR of SU(3) is given by

$$d = (1 + a_1)(1 + a_2)\left(1 + \frac{a_1 + a_2}{2}\right)$$

It is specified by two nonnegative integers a_1 and a_2. For IR [8] of SU(3), characterized by the pair of nonnegative integers ($a_1 = 1$, $a_2 = 1$), construct a YT. However, as for YTs the condition $\lambda_1 \geq \lambda_2$ for two nonnegative integers should always be satisfied, we choose $\lambda_1 = a_1 + a_2$ and $\lambda_2 = a_2$ for drawing any such tableau. Now draw a YT with two rows of boxes such that the first row contains $\lambda_1 = a_1 + a_2$ boxes while the second row contains $\lambda_2 = a_2$ boxes.

Thus,

$$8 \equiv (a_1 = 1, a_2 = 1) = (\lambda_1 = a_1 + a_2 = 2, \lambda_2 = a_2 = 1)$$

$\equiv$ 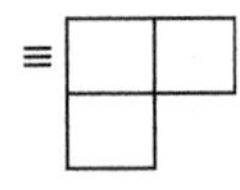

Therefore, we can write

8 x 8 = x

2. When we have drawn the two YTs for the product, we consider the second tableau and put a letter say "a" in all the boxes of the first row, and a letter, say b, in all the boxes of the second row (and so on):

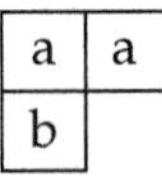

3. We enlarge the first tableau by adding to it in all possible ways those boxes from the second tableau that are marked a but taking care that the following conditions are always satisfied:
   (a) The tableau remains regular, that is, that there are always at least as many boxes in the first row as in the second, at least as many in the second row as in the third, and so on.
   (b) No two a's appear in the same column.

Then, in the first place, we get

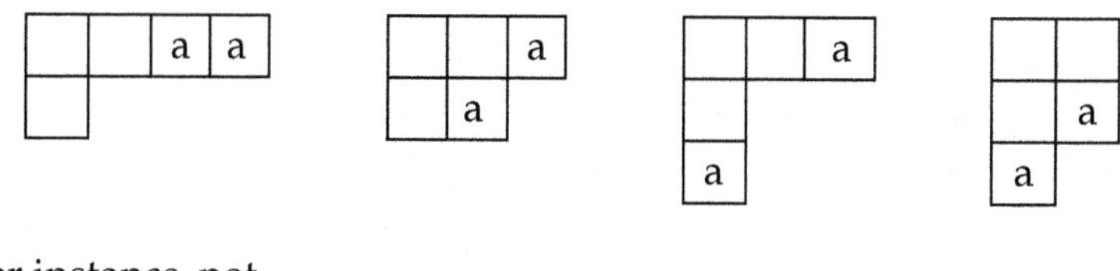

but, for instance, not

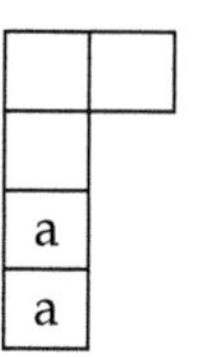

4. Observing the same rules, the b box is added to the newly formed tableaux:

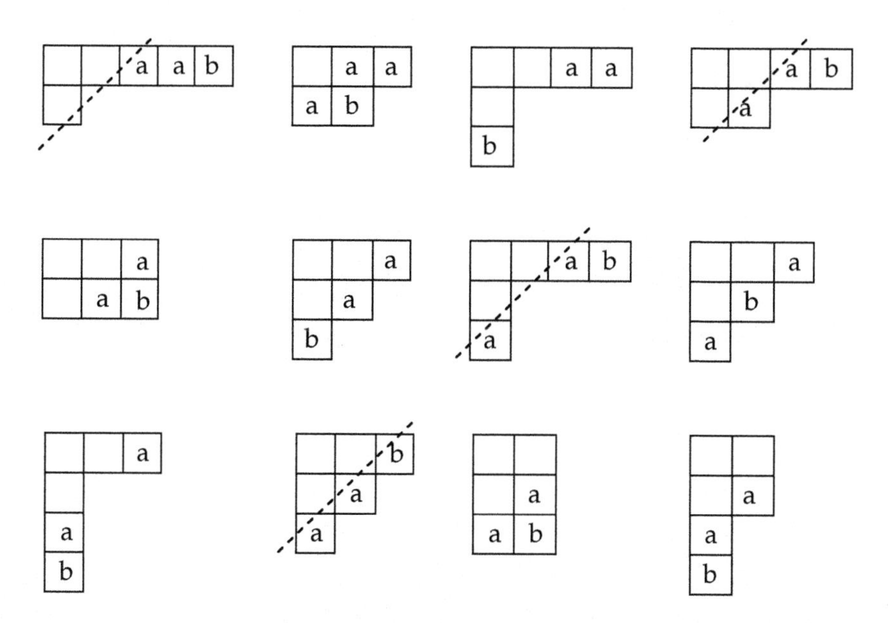

5. All of these tableaux are not to be retained. To decide which of these tableaux are to be retained, we proceed as follows:
   (a) Starting from the right of the first row of a tableau, we label the boxes of a tableau as 1,2,3,4,... in such a way that after the boxes of the first row are exhausted, the labeling is continued from the right of the second row, and so on. If, in an arbitrary set of first boxes, the number of boxes containing b is greater than the number of boxes containing a, b > a, then the tableau is rejected. Such an arrangement of boxes is known as *lattice permutation.*

(b) By virtue of this, we are left with the following eight tableaux:

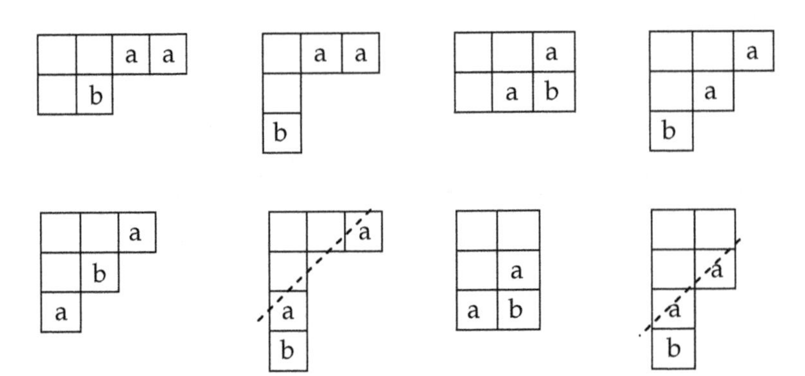

6. Any YT is rejected if the number of its one or more columns exceeding the dimensions of the group. As SU(3) is a special unitary group in a complex space of three dimensions, the tableaux with columns with four or more boxes are rejected. Hence, we are left with the following tableaux:

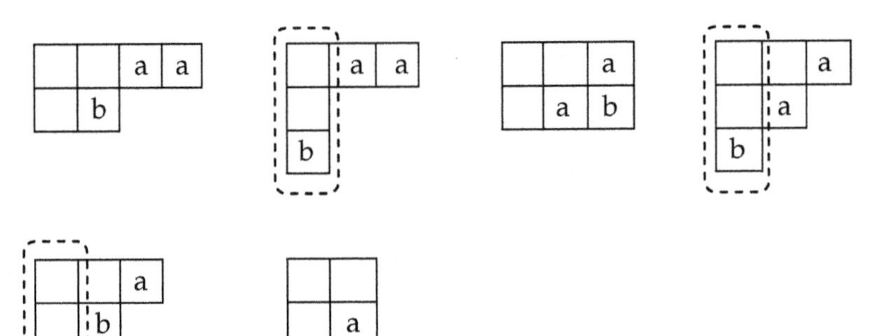

7. Any *complete column* (i.e., any column with the number of boxes equal to the dimension of the group) should be dropped. If all the columns of a YT are complete, the corresponding IR should be characterized by (0,0). This gives

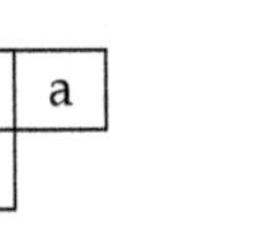

Hence,

$$\begin{aligned}
8 \times 8 &= (\lambda_1 = 4, \lambda_2 = 2) + (\lambda_1 = 3, \lambda_2 = 0) \\
&\quad + (\lambda_1 = 3, \lambda_2 = 3) + (\lambda_1 = 2, \lambda_2 = 1) \\
&\quad + (\lambda_1 = 2, \lambda_2 = 1) + (0,0) \\
&= (a_1 = 2, a_2 = 2) + (a_1 = 3, a_2 = 0) + (a_1 = 0, a_2 = 3) \\
&\quad + (a_1 = 1, a_2 = 1) + (a_1 = 1, a_2 = 1) + (0, 0) \\
&\equiv (2, 2) + (3, 0) + (0, 3) + (1, 1) + (1, 1) + (0, 0) \\
&= 27 + 10 + 10^* + 8 + 8 + 1
\end{aligned}$$

Hence, the Kronecker product of IRs [8] and [8] can be decomposed as a sum of IRs [27], [10], [10*], [8], [8], and [1].

Let us next find the IRs contained in the product of 3 and 3*, that is, in (1,0) x (0,1). For 3, we have $a_1 + a_2 = 1$, $a_2 = 0$, and for 3*, we have $a_1 + a_2 = 1$, $a_2 = 1$. Consequently,

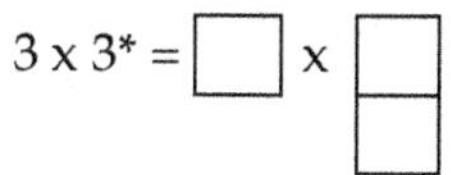

In the second tableau, writing a and b in the boxes in the first and second row, respectively, we get

a
b

Adding an a box to the first tableau in every possible way but keeping it regular, we get

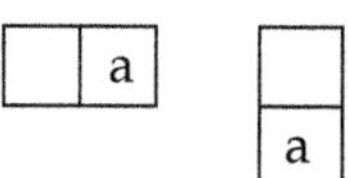

Adding a b box to these newly constructed tableaux in the same manner, we get

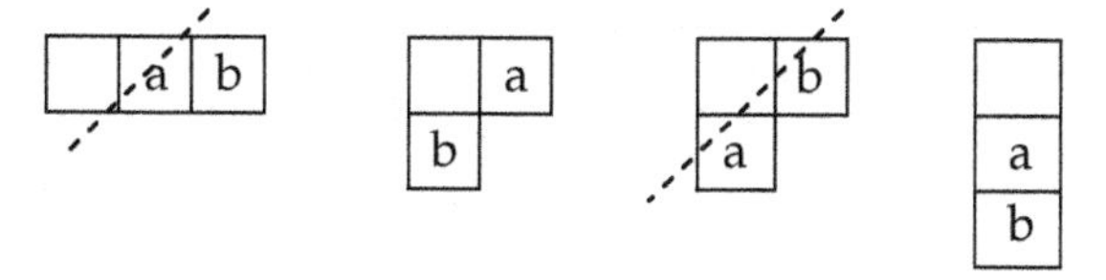

Rejecting the first and third tableaux in accordance with rule (5), we get

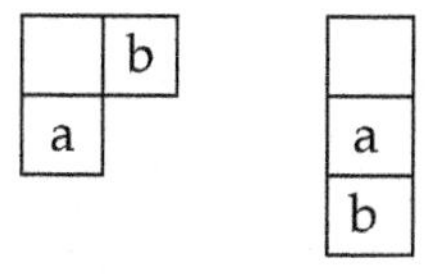

Hence,

$$\begin{aligned} 3 \times 3^* &= (\lambda_1 = a_1 + a_2 = 2, \lambda_2 = a_2 = 1) + (0, 0) \\ &= (a_1 = 1, a_2 = 1) + (0, 0) \\ &= (1, 1) + (0, 0) \\ &= 8 + 1 \end{aligned}$$

PROBLEM 3.34

By taking the product $3^* \times 3$ in this order, show that

$$3^* \times 3 = 8 + 1$$

PROBLEM 3.35

Show that the following results are valid for the irreducible representations of SU(3):

a. $3 \times 3 = 6 + 3^*$
b. $6 \times 3 = 10 + 8$
c. $8 \times 10 = 8 + 10 + 27 + 35$
d. $10 \times 10^* = 1 + 8 + 27 + 64$

The previous analysis shows that the product of two irreducible representations of a Lie group can be expressed as a sum of irreducible representations of the same group. These IRs can be ultimately used to accommodate elementary particles.

4

Symmetry, Lie Groups, and Physics

Having developed group theory to a level that it can be applied to solve the problems in high energy physics, we will now show how the symmetry principles associated with the group theoretical techniques were initially used to interpret some experimental results and make predictions.

4.1 Symmetry

What is *symmetry*? An object (or a system) is said to possess symmetry under a certain operation if after undergoing that operation it assumes an aspect indistinguishable from its initial configuration. The operation itself is called a *symmetry operation*. Symmetry is therefore related to unobservability. A symmetry operation is to be always carried about or with respect to a geometrical entity such as a point, a line, or a plane. This geometrical entity is called the *symmetry element*.

Let us illustrate the concept of symmetry with the help of a few examples.

4.1.1 Rotational Symmetry

Consider a square in a plane as shown in Figure 4.1. Let O be its center. Look at it and close your eyes. While your eyes are closed, let the square be rotated through 90° about its center. Upon opening your eyes, you cannot tell whether the square has been rotated; it looks the same. The square is said to possess *symmetry with respect to rotation* or *rotational symmetry* through 90° about its center. The point about which the symmetry operation of rotation has been performed is the symmetry element and is called the *center of symmetry*. Moreover, 90° or $2\pi/4$ is the smallest angle of rotation about the aforementioned center of symmetry that brings the square to a position indistinguishable from its initial configuration. In fact, on rotation through 90°, 180°, 270°, or 360°, four such positions appear in the course of a full revolution. Such a point of symmetry is called a *fourfold point of rotational symmetry* or a *point of fourfold rotational symmetry*. After each of these symmetry operations, the square acquires a position indistinguishable from its initial position and is said to be *invariant* under any one of the *four rotational-symmetry operations*. This fourfold rotational symmetry is a characteristic of the square.

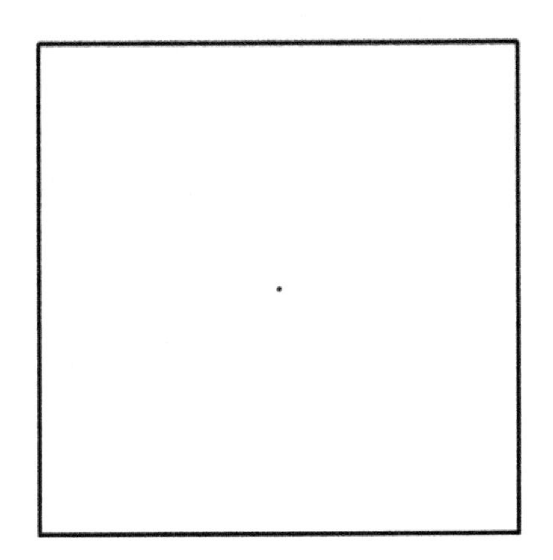

FIGURE 4.1
Rotational symmetry of a square about its center.

This is an example of a *discrete symmetry* for which the smallest value of the symmetry operation, the angle of rotation in this case, can be specified.

If the square is rotated through one complete revolution about its center, it acquires a position identical to its initial position. This operation is called the *identity operation.* A complete revolution about the symmetry element (i.e., the center O) may be attained by n successive rotations about the center, each through $2\pi/n$ radians, where n is a positive integer.

Consider again a square, a plane figure, placed horizontally in space. Let AB be a line perpendicular to the plane of the square and passing through its center (Figure 4.2). The rotation of the square through 90° about AB will bring it to a configuration indistinguishable from the initial one. The symmetry operation is rotation, and AB is the symmetry element, usually called the *line* or *axis of symmetry.*

Consider next a cube with one face horizontal as shown in Figure 4.3. Then the cube remains indistinguishable from its initial configuration if it is rotated through 90° about a vertical line AB passing through its center. The rotation through 90° is a symmetry operation, and the line AB about which the operation is carried out is a symmetry element and is called the axis of symmetry. The figure remains invariant under this fourfold axis of rotational symmetry. The rotations about the same axis mutually commute. However, rotations about different axes do not commute. In general, an object is said to have n-fold rotational symmetry if it looks the same after being rotated 360°/n degrees about a point or an axis. This is also a discrete symmetry.

A circle (or a round plate), shown in Figure 4.4, possesses *rotational symmetry* about its center that serves as the symmetry element. The circle (or the round plate) may be rotated through any angle about its center; its new configuration will always be indistinguishable from its original configuration.

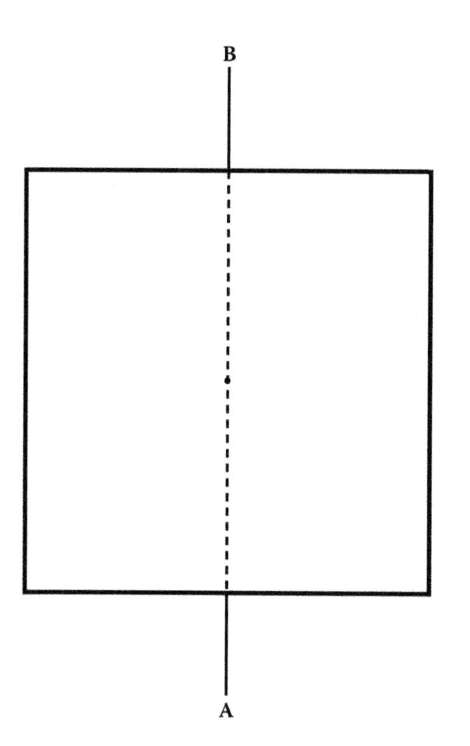

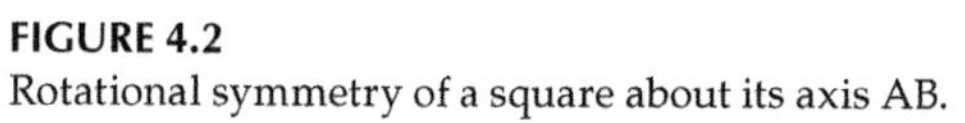

FIGURE 4.2
Rotational symmetry of a square about its axis AB.

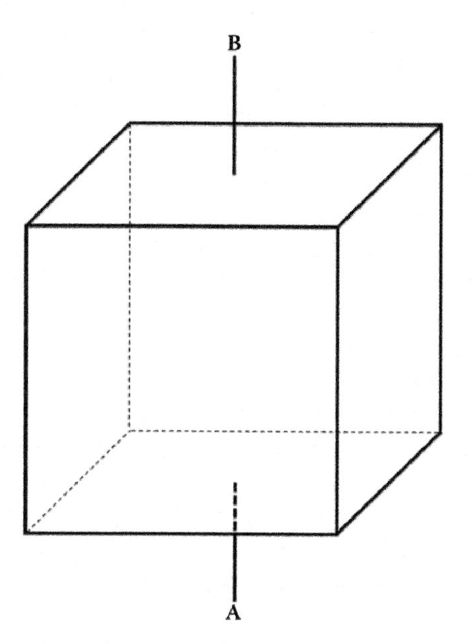

FIGURE 4.3
Rotational symmetry of a cube about its axis AB.

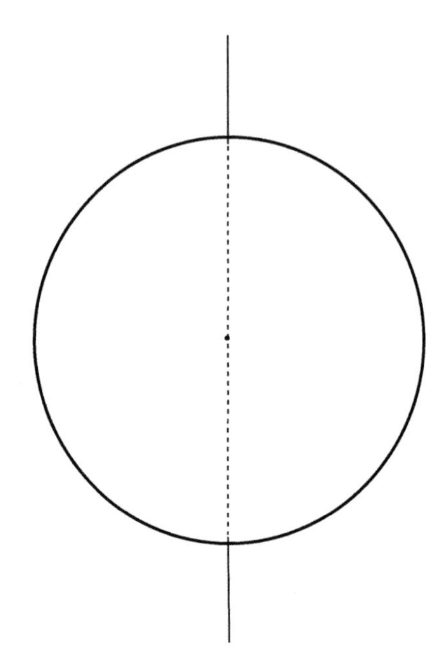

FIGURE 4.4
Rotational symmetry of a circle (or a round plate) about its center or about a perpendicular axis through its center.

However, the smallest nonzero angle of rotation cannot be specified. The symmetry is now said to be a *continuous symmetry*.

The same circle (or round plate) may be rotated about a line perpendicular to the plane of the circle (or round plate) and passing through its center. The symmetry operation is now carried out about this line, which serves as the symmetry element and is called an *axis of symmetry*.

We can look at the situation from another angle. Let us say that the round plate is kept fixed. Then the physical characteristics of the system (the round plate) associated with the x- and y-axes are the same, whatever the directions in which these axes are chosen. This leads to the conclusion that there is no absolute direction in space: the space is *isotropic*. A natural consequence of isotropy of space is that if an isolated experimental set up is rotated in space through any angle, then the results for physical characteristics are unaffected by this rotation: the experimental results are independent of the direction of the set up in space.

A circle therefore possesses rotational symmetry either about its center or about an axis passing through its center and perpendicular to its plane. This is a continuous symmetry because the smallest nonzero angle of rotation cannot be specified.

A sphere (or a spherical system) possesses *rotational symmetry* about its diameter that serves as the symmetry element. This diameter is its axis of symmetry. The sphere may be rotated through any angle about its any diameter, and its new position would always be indistinguishable from its original position. There is no smallest nonzero angle through which the sphere may be rotated (about its diameter). There are thus an infinite number of symmetry operations that can be carried out on the sphere: the symmetry is continuous.

4.1.2 Higher and Lower Symmetries

An n-fold symmetry of an object is said to be higher than its m-fold symmetry if $n > m$. Objects of high symmetries also contain lower symmetries. For instance, a square placed horizontally has fourfold symmetry about a vertical axis passing through its center. But it also has twofold symmetry about the same axis. That is, if its form remains indistinguishable after a rotation through 90° about the axis of symmetry, it also remains indistinguishable after rotation through 180° about the same axis. In general, an n-fold symmetry includes any symmetry of lower number that divides n. For instance, a hexagon has $360°/6 = 60°$ degree rotational symmetry of order 6 about the axis of symmetry and includes lower rotational symmetries of order 3 and 2.

4.1.3 Reflection/Inversion Symmetry

For a one-dimensional rod shown in Figure 4.5, if a space coordinate x is changed to $-x$ with respect to the origin O lying at the center of the rod, the configuration of the rod remains indistinguishable from its initial configuration and the rod is said to possess *reflection symmetry* about the

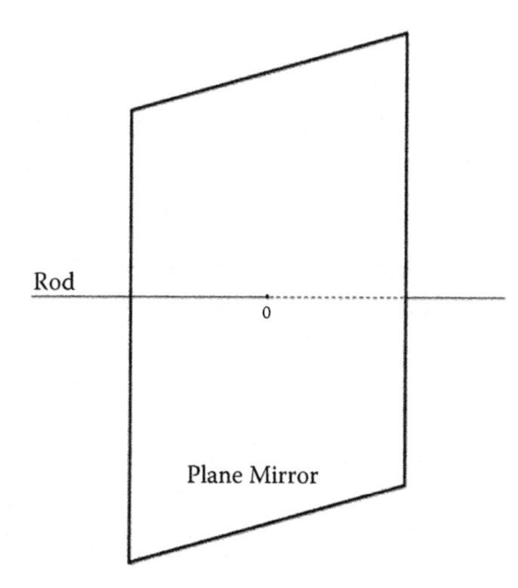

FIGURE 4.5
Reflection symmetry of a rod about its center.

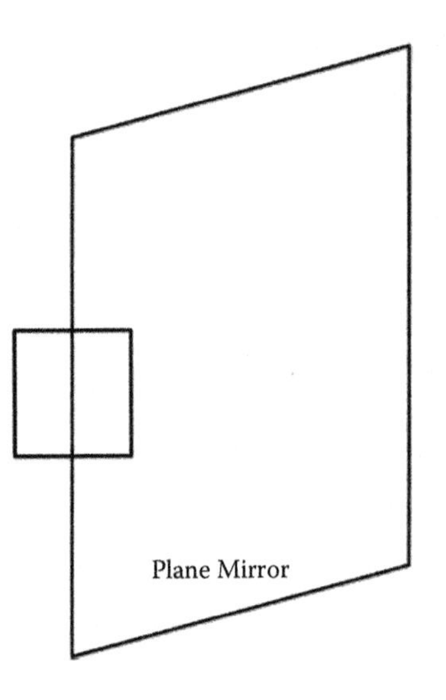

FIGURE 4.6
Reflection symmetry of a square.

symmetry element O. This is because it is equivalent to the mirror image of the rod through a plane mirror placed perpendicular to the rod and passing through O.

Let us now imagine a plane figure, such as a square like that shown in Figure 4.6, and a *double mirror* perpendicular to the plane of the square and passing through its center. Then one half of the square is the mirror image of the other half. This fact is expressed by stating that the square possesses reflection symmetry with respect to a plane mirror perpendicular to it and passing through its center. The presence of the plane mirror, the symmetry element, does not change the position of the plane figure. This symmetry is also exhibited in Figure 4.7 where one part of the butterfly is the mirror image of the other. These are examples of objects that remain invariant under reflection symmetry. In biology, it is often called *bilateral symmetry.* In these two examples, the figure or the object in the presence of the double mirror remains indistinguishable from its configuration in the absence of the mirror. Such a plane is called a *plane of reflection symmetry.* This is the only symmetry element possessed by many living organisms.

An equilateral triangle having reflection symmetry about three symmetry elements, the lines of symmetry, is shown in Figure 4.8. In general, if there exists a point in an object such that changing the space coordinates (x, y, z) of every point to (−x, −y, −z), with respect to that point serving as the origin, the resulting configuration of the object is indistinguishable from the original one, the symmetry operation is called *inversion* about or with respect to that point, and the point, serving as a symmetry element, is said to be center of symmetry or center of inversion.

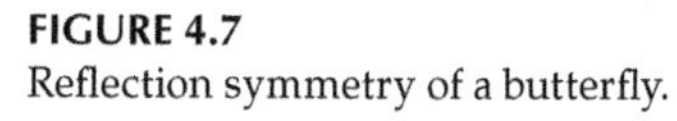
FIGURE 4.7
Reflection symmetry of a butterfly.

4.1.4 Concept of Parity

Let us now introduce the concept of *parity,* which has no classical analog. For this purpose, let us consider one-dimensional time-independent Schrödinger equation:

$$\frac{d^2\psi(x)}{dx^2}+\frac{2m}{\hbar^2}\left(E-V(x)\right)\psi(x)=0 \tag{4.1a}$$

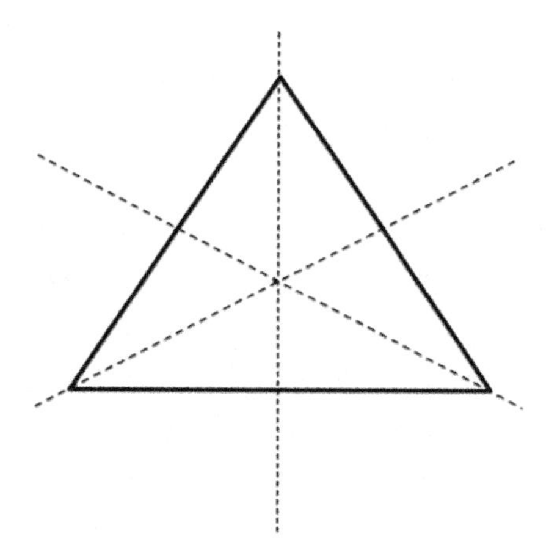

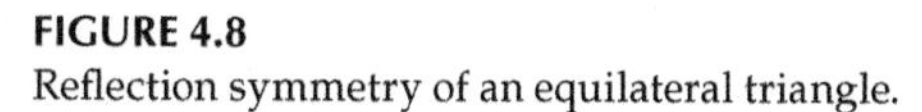
FIGURE 4.8
Reflection symmetry of an equilateral triangle.

If we now change x to −x, we get the equation for the mirror image configuration:

$$\frac{d^2\psi(-x)}{dx^2}+\frac{2m}{\hbar^2}(E-V(-x))\psi(-x)=0 \tag{4.1b}$$

If the potential energy is symmetric about x = 0, then V(−x) = V(x), and Equation (4.1b) becomes

$$\frac{d^2\psi(-x)}{dx^2}+\frac{2m}{\hbar^2}(E-V(x))\psi(-x)=0 \tag{4.2}$$

Comparing Equations (4.1a) and (4.2), we find that for the potential V(x) of a conservative system that is symmetric with respect to space coordinate x about x = 0, i.e., V(−x) = V(x), both the functions ψ(x) and ψ(−x) are the solutions of the Schrödinger wave equation. Therefore, if we denote this operation of mirror reflection by P, the plane mirror being the symmetry element, we must have

$$P\,\psi(x) = \psi(-x)$$

Operating again by P, we get

$$P^2\,\psi(x) = P\,\psi(-x) = \psi(x)$$

The last step has been taken because a double reflection leads back to the original configuration. This is an eigenvalue equation and shows that the eigenvalue of the operator P^2 is 1. If we use the same symbol for the operator as well as its eigenvalue, we may write $P^2 = 1$. This yields $P = \pm 1$. The operator P is called the parity operator, and its eigenvalue is called the parity. The parity of a wave function ψ(x) is said to be even or odd according to if its value

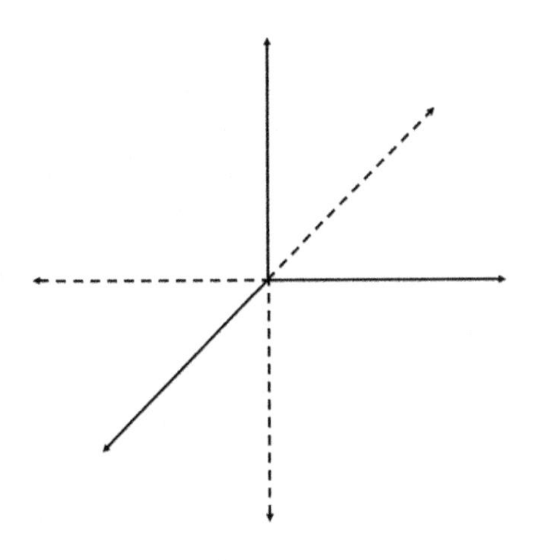

FIGURE 4.9
Transformation of a right-handed coordinate system into a left-handed coordinate system.

is +1 or −1. Hence, the solution of the time-independent Schrödinger wave equation with symmetrical potential $V(-x) = V(x)$ is either even or odd under a change of sign in the space coordinate x: the even solutions are said to have even parity, and the odd solutions have odd parity. The usual terminology is that if the mirror image is also realized in nature with equal probability, parity is conserved; otherwise it is violated. In 1957, parity was found to be violated in weak interactions.

It may be pointed out that under parity transformation (i.e., $x \to -x$, $y \to -y$, $z \to -z$) a right-handed coordinate system is transformed into a left-handed one or vice versa. It was thought that nature should not care whether a coordinate system is right-handed or left-handed. But parity violation in weak interactions showed that it did. Nature is not symmetric under mirror reflections. This fact is expressed by stating that for weak interactions nature has only a "left hand" and no "right hand."

REMARK

Wolfgang Pauli, a Nobel Laureate in physics, was a versatile genius with a mastery over satire. He did not believe that parity would be violated (in weak interactions) and expressed his opinion by stating that God cannot be left-handed. However, when parity violation in weak interactions was confirmed experimentally, he wrote to Victor Weisskopf[1]: "It is good that I did not make a bet. It would have resulted in a heavy loss of money (which I cannot afford). I did make a fool of myself, however (which I think I can afford to do):... What shocks me is not the fact that God is just left-handed but the fact that in spite of this He exhibits Himself as left/right symmetric when He expresses Himself strongly."

4.1.5 Multiple Symmetries

An object that possesses more than one type of symmetry has *multiple symmetries*. For instance, consider a square, a plane figure. It has rotational symmetry through 90° or an integral multiple of it about its center or about a line perpendicular to the square and passing through its center as it can be rotated through 90° or an integral multiple thereof to a configuration indistinguishable from its initial configuration. But it is also mirror symmetric about each of four lines passing through its center as shown in Figure 4.10.

4.1.6 Combination of Symmetry Operations

Symmetry operations can also be combined. Suppose that a symmetry operation can be performed on an object, say, by rotating it through 360°/n, where n is a specific integer, about an axis of rotation perpendicular to the figure and passing through its center. The operation is called *proper rotation*, and the axis is

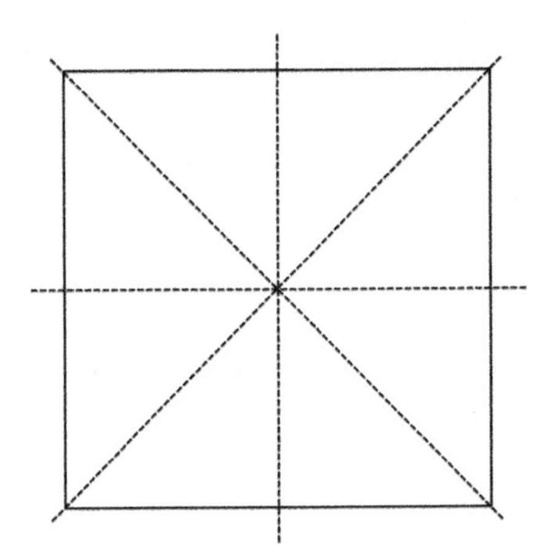

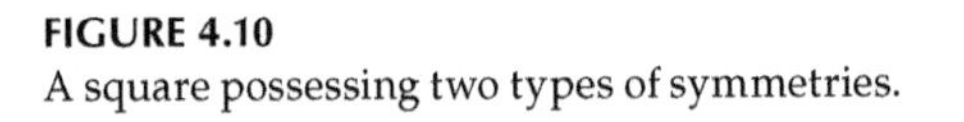

FIGURE 4.10
A square possessing two types of symmetries.

said to be the *proper axis of rotation*. If, followed by a reflection through the plane perpendicular to the rotation axis, the resulting configuration is still indistinguishable from the original, the two symmetries are said to be combined. The axis is then called an n-fold *improper axis of rotation*. The object is said to possess *improper rotational symmetry* about an n-fold improper rotation axis.

PROBLEM 4.1
Show that a combination of a plane mirror symmetry with an even-fold axis yields plane mirror symmetries created by rotation, plus another set of halfway between them.

4.1.7 Translational Symmetry in Space

Next, consider an *infinite* pattern repeated after a length L as shown in Figure 4.11. If the eyes are closed and someone displaces the pattern longitudinally in its plane through a distance which is an integral multiple of L, it is

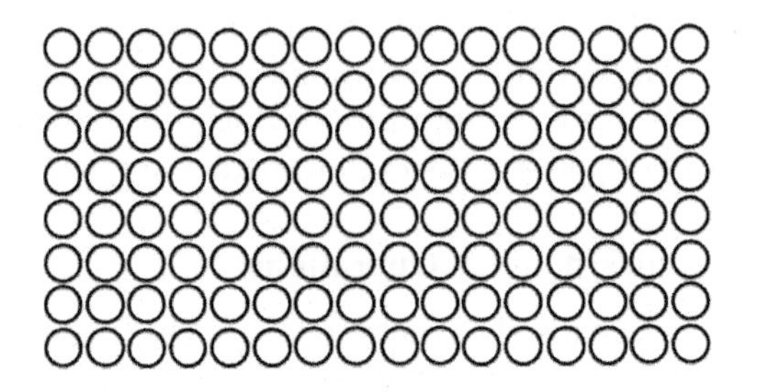

FIGURE 4.11
Translational symmetry in space.

not possible to determine whether, during the interval the eyes were closed, the pattern has been displaced or not. Such a figure is said to exhibit *translational symmetry* with respect to a longitudinal displacement, the displacement line serving as the symmetry element, or is said to be *invariant under translational symmetry*. This shows that there is no absolute position in space: the space is *homogeneous*. The translation of an isolated experimental set up in space will not change the results of the experiment.

4.1.8 Time-Reversal Symmetry

Time reversal just means replacing t by $-t$, thus inverting the direction of flow of time. The time-reversed process is defined as being obtained by changing t to $-t$. In fact, $t \to -t$ leads to the linear momentum $\mathbf{p}$ ($= m\, d\mathbf{r}/dt$)$\to -\mathbf{p}$ and the angular momentum $\mathbf{L}$ ($= \mathbf{r} \wedge \mathbf{p}$) $\to -\mathbf{L}$. Thus, if we consider the reaction

$$a + b \to c + d$$

the time-reversed process is

$$c + d \to a + b$$

In this process, the linear and angular momenta of all the particles are reversed. Time-reversal invariance requires that these two processes should occur with the same probability. Hence, time-reversal invariance means that the two processes, one obtained from the other by reversing the directions of linear and angular momenta have equal reaction rates.

The linear transformation (called time reversal)

$$(x' = x,\ y' = y,\ z' = z),\ t' = -t$$

also satisfies the equation

$$x'^2 + y'^2 + z'^2 - c^2 t'^2 = x^2 + y^2 + z^2 - c^2 t^2$$

and is therefore a Lorentz transformation. The determinant of the transformation matrix is -1; it is hence an improper Lorentz transformation. If time-reversed transformation is a valid operation, say, for the motion of an object, it will be equally permissible to carry out the motion of this object in the reverse order, like a movie played backward. If time reversal is violated, such a motion of the object should not be possible. In fact, the initial conditions are usually such that it is possible to make a difference between forward and backward motions. The validity of the time-reversal invariance is in doubt at present.

Let us now examine the behavior of Newton's second law of motion under time reversal. The law is expressed by the differential equation

$$\mathbf{F} = \frac{d\mathbf{p}}{dt} = \frac{d}{dt}\left(m\frac{d\mathbf{r}}{dt}\right)$$

For a conservative system, the law is invariant under the time-reversal transformation $t \rightarrow -t$, as time is occurring in the second power.

Let us next study the effect of time reversal in quantum mechanics. Consider a system whose behavior is governed by the time-dependent Schrödinger wave equation:

$$H\psi(\mathbf{r},t) = i\hbar\frac{\partial}{\partial t}\psi(\mathbf{r},t)$$

If we change t to −t, the previous equation takes the form

$$H\psi(\mathbf{r},-t) = -i\hbar\frac{\partial}{\partial t}\psi(\mathbf{r},-t)$$

This shows that the Schrödinger wave equation is not invariant under the time-reversal transformation. If, however, we take the complex conjugate of each side, we get

$$H\psi(\mathbf{r},-t) = i\hbar\frac{\partial}{\partial t}\psi^*(\mathbf{r},-t)$$

This is the Schrödinger wave equation and projects the fact that $\psi^*(\mathbf{r},-t)$ is also a solution of the Schrödinger wave equation. Thus, $\psi(\mathbf{r},t)$ and $\psi^*(\mathbf{r},-t)$, the latter being the complex conjugate of the time-reversed solution, are both solutions of the same Schrödinger wave equation. However, in quantum mechanics, the measurement of a physical quantity always involves $\psi^*\psi$. For a conservative system, we have

$$\psi(\mathbf{r},t) = \psi(\mathbf{r}) \exp(-iEt)$$

and

$$\psi^*(\mathbf{r},t) = \psi^*(\mathbf{r}) \exp(iEt)$$

This yields

$$\psi^*(\mathbf{r},t)\,\psi(\mathbf{r},t) = \psi^*(\mathbf{r})\,\psi(\mathbf{r})$$

For the other solution, we have

$$\psi(\mathbf{r},-t) = \psi(\mathbf{r}) \exp(iEt)$$

and

$$\psi^*(\mathbf{r},-t) = \psi^*(\mathbf{r}) \exp(-iEt)$$

This yields

$$\psi^*(\mathbf{r},-t)\,\psi(\mathbf{r},-t) = \psi^*(\mathbf{r})\,\psi(\mathbf{r})$$

Hence, for a conservative system, both the solutions will give the same physical result.

The previously provided analysis shows that the origin of time is arbitrary. This is called the *homogeneity of time*. The repetition of an experiment at any time will yield the same result provided the setup is isolated. In fact, laws of physics are symmetric with respect to translations in space and time as well as rotations in space. This means that the translation of an isolated physical system or its rotation through any angle about any axis will not change the experimental results, no matter when the experiment is performed.

Symmetries such as those that occur under rotation, reflection, and translation and whose description involves space–time coordinates are called *external* or *geometrical* or *space–time symmetries*. However, other types of symmetries are independent of space and time coordinates and are called *internal* or *dynamical symmetries*. So far, we have considered only external symmetries. Although several types of internal symmetries exist, we will consider only charge conjugation here. We do that next.

4.1.9 Charge Conjugation

Consider a process

$$\mathrm{a} + \mathrm{b} \rightarrow \mathrm{c} + \mathrm{d}$$

A new process is obtained by changing particles to antiparticles and vice versa. This new process is

$$\bar{\mathrm{a}} + \bar{\mathrm{b}} \rightarrow \bar{\mathrm{c}} + \bar{\mathrm{d}}$$

It is said to have been obtained by the *charge conjugation* of the initial process; it is the replacement of each particle by its antiparticle and vice versa. If the probability of occurrence of both the processes is the same, we say that charge conjugation is conserved; otherwise, it is violated. The charge conjugation operation is denoted by C. It is conserved in strong and electromagnetic interactions but breaks down for weak interactions. Thus, parity and charge conjugation are both violated in weak interactions. It was thought for some time that the combined operation CP was conserved in weak interactions. However it was found in 1964 that CP was also violated in weak interactions. It has been proved in relativistic quantum field theory that under very weak conditions, CPT, the combined operation of C, the charge conjugation, P, the parity, and T, the time reversal, is an exact symmetry of any interaction. This process, obtained from a given process by applying C, P, and T, should occur with the same probability as the initial process. In fact, if experiments show that CPT symmetry is not exact, then the field theory would have to be developed anew. Since P and C symmetries are broken in weak interactions and CP is also broken, T symmetry must break down if CPT is an exact symmetry. The predictions of CPT—such as that the mass and lifetime of any antiparticle must be exactly the same as those of its particle—have been verified to such an accuracy that there is a little chance of CPT symmetry not being exact.

It has been shown in books on classical mechanics that there is a connection between symmetries and conservation laws. For systems that can be described by classical Lagrangians, symmetry of the Lagrangian under space translations, space rotations, and time translations implies the conservation of linear momentum, angular momentum, and energy, respectively. The same is valid in quantum mechanics. It may be pointed out that a conservation law is exactly valid under all circumstances only if the corresponding symmetry is exact.

Let us next consider mathematical expressions or equations as our objects. If an expression (equation) does not change in form under a certain transformation, it is said to be invariant (covariant) under that transformation. However, it is customary to use the term invariant even in the case of equations. Now Newton's equations of motion are covariant under a Galilean transformation, while Maxwell's equations are Lorentz covariant, as they retain their respective forms under these transformations. Physically, such a covariance shows that some sort of symmetry is possessed by these equations: in fact, Newton's equations retain their form in going from one inertial frame of reference to another when the frames are considered to be connected by Galilean transformations, whereas Maxwell's equations are form-invariant in going from one inertial frame to another when these frames are considered to be connected by Lorentz transformations.

The expression $E^2 + H^2$, where E and H are the magnitudes of the electric and magnetic fields, respectively, is invariant under Lorentz transformations (i.e., relativistically invariant), since this expression does not change in going from one inertial frame of reference to another.

In field theories, it is the Lagrangian density $\mathcal{L}$ (usually referred to as the Lagrangian) of a physical system that becomes the basis of all analysis. This Lagrangian has to be constructed in such a way that in a relativistic field theory it is invariant under Lorentz transformations. But it can also be invariant under some other transformation. For instance, the Lagrangian density for weak interactions is invariant under Lorentz transformations as well as chirality (from the Greek word for hand) transformation. The invariance of the Lagrangian (or ultimately the covariance of its equations of motion) under a certain transformation expresses a symmetry of the physical system. The invariance under Lorentz transformation means that the expression for $\mathcal{L}$ does not change its form in going from one inertial frame of reference to another. What does the invariance under chirality transformation imply? The mathematical concept is there, but the corresponding symmetry occurring in the physical theory cannot be visualized.

Symmetry may be broken either explicitly or spontaneously. If we add one or more terms to the Lagrangian density so that it does not remain invariant

FIGURE 4.12
Rotational symmetry of a flower.

under a symmetry group, we say that the symmetry has been explicitly broken. On the other hand, under certain conditions, although the Lagrangian density of a physical system is symmetric under a certain group of transformations, the ground state of the system is not invariant under that very group. This fact is expressed by stating that there has been a *spontaneous symmetry breaking* (SSB).

The set of symmetry operations of any given object forms a group under successive applications of operations. For instance, the set of all symmetry operations of 72° rotations of the flower about its center, as shown in Figure 4.12, forms a group. This is a finite group. The set of all rotations of a round plate about its center forms an infinite group under successive application of rotations. Group theory is therefore the most appropriate technique for studying the symmetries of physical systems.

The domain of study of symmetries extends from science to arts: the realm of symmetries contains not only geometry, physics, chemistry, and biology but also architecture and art and includes the wonderful patterns designed by nature. However, to a physicist, symmetry has significance even beyond its aesthetic qualities.

We may point out that, as a consequence of the invariance of physical laws under Lorentz transformations as introduced in the theory of special relativity by Einstein, the emphasis in physical theories was shifted to symmetries of nature considered more fundamental than anything else. Since then these have played a significant role in theoretical physics. In fact, the importance of symmetries in physics stems from the nonavailability of differential equations and mathematical tools required for the solution of equations occurring in various theories. The symmetry of a system under a certain transformation or operation gives an opportunity to obtain some information about the system and its characteristics even when the corresponding dynamical equations or their exact mathematical solutions are not available. However, the discovery of parity violation in weak interactions showed that symmetries could be violated, and hence experiment is the ultimate test for checking the validity of any symmetry.

4.1.10 Symmetry Groups and Physics

Let us now consider an event noticed by two observers in two different inertial frames of reference, one moving relative to the other with uniform velocity v, and along the common x-axis. According to Newtonian mechanics, the space coordinates x, y, z of the event and the time t at which it has occurred as measured by one observer can be compared with the corresponding space and time measurements x′, y′, z′, t′ of the same event by the other observer through Galilean transformations

$$x' = x - vt$$

$$y' = y$$

$$z' = z$$

$$t' = t$$

The last equation shows that in Newtonian mechanics, time is an absolute quantity. In matrix form, this set of homogeneous nonsingular linear transformations may be written as

$$x' = Ax$$

PROBLEM 4.2

Write down the expression for the transformation matrix A, and show that the determinant of the matrix is 1.

Since the determinant of the matrix is different from 0, the inverse of A exists. Moreover, the real parameter v can vary continuously. Hence,

$$x' = Ax, \qquad |A| = 1$$

the Galilean transformations form a continuous group of homogeneous linear transformations under successive application of transformations. Explicit calculations show that under these transformations, which form a group, the second law of motion,

$$\mathbf{F} = \frac{d\mathbf{p}}{dt} = m\frac{d^2\mathbf{r}}{dt^2}$$

does not change its form in going from one inertial frame to another. This fact is also expressed by stating that Newton's law of motion is invariant under continuous *group of Galilean transformations.* This group is therefore a symmetry group for the law of motion in classical mechanics.

Let us next examine the situation from the point of view of special relativity. One of the basic postulates of Einstein's theory of special relativity is that a law of physics should have the same form in all inertial frames of reference. That is as far as the mathematical expressions for the laws of physics are concerned, they do not change in form in going from one inertial frame of reference to another. But, according to special relativity, the space and time coordinates in two inertial frames of reference are connected through Lorentz transformations and not through Galilean transformations. For standard Lorentz transformations, the corresponding equations are

$$x' = \gamma(x - vt)$$

$$y' = y$$

$$z' = z$$

$$t' = \gamma\left(t - \frac{v}{c^2}x\right)$$

or

$$x_1' = \gamma\left(x_1 + \frac{iv}{c}x_4\right)$$

$$x_2' = x_2$$

$$x_3' = x_3$$

$$x_4' = \gamma\left(-\frac{iv}{c}x_1 + x_4\right)$$

where x_1, x_2, and x_3 stand for x, y, and z, respectively, and x_4 is ict ($i = \sqrt{-1}$, c = velocity of light, t = time). These equations can be written in matrix form as

$$x' = Ax, \quad |A| \neq 0$$

PROBLEM 4.3

Write down the expression for the transformation matrix A and find the value of its determinant.

This set of equations represents one transformation that connects the space–time coordinates x, y, z, and t of an event in one inertial frame to the space–time coordinates x′, y′, z′, and t′ of the same event as measured in the other inertial frame. Since space and time coordinates in two inertial frames are connected through standard Lorentz transformation, the laws of physics should remain invariant under these transformations. It is found by explicit calculations that Newton's law of motion changes under Lorentz transformation, and is, therefore, not a correct law. It is believed that electromagnetic phenomena are governed by Maxwell's equations. If this is true, then according to special relativity, Maxwell's equations should remain covariant under standard Lorentz transformations: they will have the same form whether we are considering an unprimed or a primed frame of reference. But the infinite set of standard Lorentz transformations that are nonsingular homogeneous linear transformations forms a continuous group under successive application of transformations, the real parameter v varying continuously. Hence, Maxwell's equations are covariant under the *group of standard Lorentz transformations*. As Maxwell's equations are known, this can be verified by actual calculations. The group of standard Lorentz transformations is a symmetry group for Maxwell's equations. However, for weak and strong interactions, the laws of physics are not known. But whatever be the laws, according to special relativity, they must have the same form in all inertial frames of reference; that is, they must be invariant under the group of standard Lorentz transformations. Explicit verification is not possible in these cases.

But it is not essential that laws of physics should be invariant only under the group of standard Lorentz transformations. These may also be invariant under some other group of transformations, say, SU(3), SU(2) × U(1). If the expressions for the laws are not known, we cannot verify this directly. But by taking these groups as symmetry groups for, say, strong interactions (SU(3)) and for weak interactions (SU(2) × U(1)), we may derive certain results and put them to experimental test. The extent of the agreement between the predictions and the experiment will reflect whether our choice of a particular symmetry group for any interaction is exactly correct, approximately valid, or incorrect. If the agreement between theory and experiment is not satisfactory, the symmetry group is changed, and new results obtained are put to experimental test, and so forth.

It may be stated that once we have chosen a symmetry group, we can determine its generators. These can be used to calculate its roots. Then, by making use of these generators and roots, we can compute the weights of any irreducible representation (IR) of the group, and the number of weights depends on the dimension of the irreducible representation. The components of each of these weights are, by definition, the eigenvalues of the mutually commuting Hermitian generators $H_1, H_2, \ldots, H_\ell$ corresponding to simultaneous set of eigenvectors. These mutually commuting Hermitian generators are assumed to be related to operators representing observables and thus can be used to directly notice some characteristics of the particles.

We will now define the Casimir operators whose eigenvalues can be used to label any irreducible representation of a Lie group instead of the nonnegative integers required to label it by specifying its dimensions.

4.2 Casimir Operators

An operator that is a nonlinear function of all the generators of a group and commutes with each of them is called a *Casimir* or an *invariant operator.* It is usually denoted by C. It has been shown by G. Racah that for a semisimple group of rank ℓ, one can construct ℓ Casimir operators.[2,3] Since C commutes with all the generators of the group, by virtue of Schur's lemma, its significance stems from the fact that it must be a constant multiple of identity. The eigenvalues of these operators uniquely characterize every irreducible representation of the group.

The three-parameter rotation group O(3) of rank ℓ has three linearly independent generators and one Casimir operator. Incidentally, it may be pointed out that the groups SU(2) and O(3) have the same Lie algebra but are not isomorphic. There exists a homomorphic (two-to-one) mapping of SU(2) onto O(3). Let J_x, J_y, and J_z be three generators of O(3). Then $J^2 = J_x^2 + J_y^2 + J_z^2$, a nonlinear function of the three generators, commutes with each one of them and is therefore a Casimir operator. The members of each IR of O(3) have the same eigenvalue $j' = j(j + 1)$ of J^2 but can be distinguished by quantum number m, the eigenvalue of J_z, say.

The eight-parameter group SU(3) of rank 2 has eight linearly independent generators and two Casimir operators, say C_1 and C_2. The dimension d of any one of its IRs is given by

$$d = \left(1 + \frac{(a+b)}{2}\right)(1+a)(1+b)$$

where a and b are nonnegative integers. Each IR of SU(3) is, therefore, specified by two nonnegative integers a and b. It can be shown that the eigenvalues of the Casimir operators for SU(3) are given by

$$C_1(a, b) = (a^2 + b^2 + ab + 3(a + b))/3$$

$$C_2(a, b) = (a - b)(2a + b + 3)(2b + a + 3)/18$$

Hence, the IRs of SU(3) can be characterized, instead of the parameters a and b, by the eigenvalues of the Casimir operators C_1 and C_2. The expressions for the eigenvalues of these two operators are, respectively, symmetric and antisymmetric with respect to the interchange of a and b. Thus, the sign of $C_2(a,b)$ distinguishes a representation D(a,b) from its complex conjugate representation $D^*(a,b)$.

Theorem 4.1

Casimir operators of a semisimple Lie group are not unique.

PROOF

Casimir operators $C_1, C_2, \ldots, C_\ell$ of a semisimple Lie group of rank ℓ commute, by definition, with the generators L_A of the group. Therefore, their linear combinations will also commute with L_A and for that reason can serve as Casimir operators of the group. Hence, Casimir operators are not unique. However, it must be noted that the number of linearly independent Casimir operators of a Lie group of rank ℓ is always ℓ. ■

PROBLEM 4.4
Show that Casimir operators of a unitary semisimple Lie group can always be *chosen* Hermitian.

4.3 Symmetry Group and Unitary Symmetry

Before considering the role symmetry can play in understanding the physics of a system, we define unitary symmetry.

Consider a system. We know that if the Hamiltonian or the Lagrangian or the equations of motion of a given system remain invariant under a group of operators or transformations, then this group is called a symmetry group of the system. For strong interactions, the Hamiltonian of a physical system is not known. Therefore, we cannot verify directly whether it is invariant under a particular group of transformations or not. However, by assuming a particular symmetry group for strong interactions, we can predict certain results that can be tested experimentally, thus either justifying or contradicting our assumption about the symmetry group. If the Hamiltonian of a physical system is invariant under a group of unitary transformations, then the system is said to possess *unitary symmetry.*

4.4 Symmetry and Physics

We will now see how much we can learn about the physics of a system from its symmetry properties alone. Let us consider a system that possesses only one symmetry S_α described by a unitary semisimple Lie group. For instance, a system may have spherical symmetry described by the rotation group SO(3) ≡ R(3). We have imposed this restriction of a single symmetry to keep the calculations simple. This restriction can be removed to extend the analysis to include systems possessing several symmetries at the same time. Let us derive some characteristics of the system possessing a single symmetry.

Theorem 4.2

Show that if a system possesses a single symmetry S_α described by a symmetry group G_α, then the Hermitian generators L_A of the symmetry group commute with the Hamiltonian operator H of the system so that $[L_A, H] = 0$ and hence the physical quantities corresponding to the Hermitian generator L_A are conserved.

PROOF

Let U(α) be unitary operators of a representation of a group G_α. Under the action of a unitary operator, the state vectors of the system are transformed as $|\Psi'\rangle = U(\alpha)|\Psi\rangle$ whereas the new Hamiltonian is given by

$$H' = U(\alpha)HU^{-1}(\alpha) \tag{4.3}$$

The system possesses a symmetry described by G_α and is therefore invariant under this transformation; that is, its Hamiltonian operator is unchanged:

$$H' = H$$

Writing H for H′ in Equation (4.3), we get

$$H = U(\alpha)HU^{-1}(\alpha)$$

or

$$HU(\alpha) = U(\alpha)H$$

that is, H commutes with U(α) for all α. Therefore,

$$H\frac{\partial U(\alpha)}{\partial \alpha_A}\bigg|_{\alpha=0} = \frac{\partial U(\alpha)}{\partial \alpha_A H}\bigg|_{\alpha=0}$$

or

$$HL_\alpha = L_\alpha H$$

or

$$[L_\alpha, H] = 0$$

Hence, the physical quantities represented by the Hermitian generators L_A of the symmetry group are conserved.

For a group of rank ℓ, only ℓ of these operators L_A mutually commute. Hence, only ℓ of the physical quantities can be simultaneously conserved. ■

Example 4.1

The spherical symmetry possessed by a system is described by the rotation group SO(3): $R(\boldsymbol{\theta}) = \exp(i\boldsymbol{\theta}\cdot\mathbf{J})$. The Hermitian generators of this group are J_1, J_2, and J_3. The physical quantities corresponding to them are three components of the angular momentum J. Since J_1, J_2, and J_3 do not commute, only one of these components is conserved at a time.

Theorem 4.3

If a system possesses a single symmetry S_α, say rotational symmetry, described by the group G_α, then the Casimir operators C_i of the group commute with the Hamiltonian H of the system so that we may write $[H, C_i] = 0$, and hence the physical quantities corresponding to ℓ Hermitian operators C_i are conserved simultaneously.

PROOF

We have proved already that if H is the Hamiltonian of a system and G_α is a symmetry group, then

$$[H, L_A] = 0$$

where L_A are the generators of the group. Therefore, for any function $f(L_A)$ of L_A, we have

$$[H, f(L_A)] = 0$$

This means

$$[H, C_i] = 0$$

because C_i, the Casimir operators, are merely nonlinear functions of L_A.

Hence, if the operator C_i does not explicitly depend upon time, then, as $[H, C_i] = 0$, the physical quantity corresponding to the operator C_i is conserved. Since C's mutually commute, the ℓ physical quantities, the eigenvalues of C_i's, are conserved simultaneously. The ℓ simultaneous eigenvalues of the invariant operators C_i of the group are thus *good quantum numbers* of the system. These eigenvalues of C_i can be used to label the multiplets of the group G_α uniquely. ■

Theorem 4.4

If a system possesses a symmetry S_α described by a symmetry group G_α, then all the transitions from one multiplet of G_α to another are absolutely forbidden.

PROOF

Let m and m′ be two multiplets (which do not contain any irreducible invariant subspace) of the symmetry group G_α. Let C_i be the invariant operators of this group and j_i and j'_i ($\neq j_i$) their eigenvalues labeling uniquely m and m′. Then

$$C_i|j_i> = j_i|j_i>$$

$$C_i|j'_i> = j_i'|j_i'>$$

If H is the Hamiltonian of the system, then, as we have proved,

$$[C_i, H] = 0$$

or

$$C_iH - HC_i = 0$$

Operating from the left by $<j_i'|$ and from the right by $|j_i>$, we get

$$<j'_i|C_iH - HC_i|j_i> = 0$$

or

$$<j'_i|C_iH|j_i> - <j'_i|HC_i|j_i> = 0$$

or

$$j'_i<j'_i|H|j_i> - j_i<j'_i|H|j_i> = 0$$

or

$$(j_i' - j_i)<j'_i|H|j_i> = 0$$

or

$$<j_i'|H|j_i> = 0, \text{ because } j'_i \neq j_i$$

This proves the theorem. ■

4.5 Group Theory and Elementary Particles

Let us assume that strong interactions are invariant under SU(3), the special unitary group in a three-dimensional complex space. This group is then the symmetry group for strong interactions. We further assume that the strongly interacting particles, called hadrons, are the basic states in the irreducible representations of the symmetry group SU(3) so that a d-dimensional irreducible representation can accommodate d particles. However, elementary particles cannot be arbitrarily accommodated in various irreducible representations. A d-dimensional irreducible representation of a simple Lie group of rank ℓ has d weight vectors, each vector having ℓ components. Since H_i are Hermitian operators, their eigenvalues are real. But these eigenvalues are the components of a weight vector. Therefore, we can associate every component of a weight vector with an observable. Of course, the way the various characteristics of elementary particles are associated with the components of the weights of an irreducible representation of a group determines the model of elementary particles.

Let us now consider two models based on SU(3), a group of rank 2 and therefore having two commuting Hermitian generators H_1 and H_2. We start with a *fundamental representation* 3 of SU(3). It can accommodate three particles.

Let us find out which particles can be accommodated in this irreducible representation of SU(3). For the IR $D^{(3)}(1,0)$, the three weights, as already determined, are

$$\mathbf{m}(1)=\left(\frac{1}{2\sqrt{3}},\frac{1}{6}\right)$$

$$\mathbf{m}(2)=\left(-\frac{1}{2\sqrt{3}},\frac{1}{6}\right)$$

$$\mathbf{m}(3)=\left(0,-\frac{1}{3}\right)$$

The first and second components of each weight are, respectively, the eigenvalues of two commuting Hermitian generators H_1 and H_2 of SU(3). Therefore, if H_1 and H_2 are associated with the operators representing some observables, the eigenvalues of H_1 and H_2 will be associated with the eigenvalues of those operators, and these eigenvalues will give the measured values of observables. In 1956, S. Sakata[4] proposed a new model according to which all hadrons could be constructed by proton, neutron, Λ-particle and their antiparticles. It was then assumed that the operators H_1 and H_2 are associated with the operator I_3, the third component of isospin, and the operator Y, the hypercharge, by the relations

$$\sqrt{3}H_1 = I_3$$

and

$$2H_2 + \frac{2}{3} = Y$$

The model based on these relations is called the *Sakata model*. Let us make explicit calculations. Consider the weight **m**(1). The two components of the weight are the eigenvalues of the commuting Hermitian generators H_1 and H_2 corresponding to a simultaneous eigenket $|m(1)>$ of these generators. Therefore, we may write

$$I_3 \ |m(1)> = \sqrt{3}\ H_1 \ |m(1)> = \sqrt{3}\ m_1 \ |m(1)> = \frac{1}{2} \ |m(1)>$$

Then 1/2 is the eigenvalue of the operator I_3, the third component of the isospin operator.

Similarly,

$$Y|m(1)\rangle = \left(2H_2 + \frac{2}{3}\right)|m(1)\rangle = 2m_2\,|m(1)\rangle + \frac{2}{3}|m(1)\rangle$$

$$= \left(2 \cdot \frac{1}{6} + \frac{2}{3}\right)|m(1)\rangle = |m(1)\rangle$$

If we denote the eigenvalues of the operators I_3 and Y by the same letters, we may write

$$I_3 = \frac{1}{2}$$

and

$$Y = 1$$

That is, the particle associated with this weight must have $I_3 = 1/2$ and $Y = 1$. The value of the charge Q on this particle is obtained by using the Gell-Mann–Nishijima formula,

$$Q = I_3 + \frac{Y}{2}$$

This yields $Q = 1$. The proton has got all these characteristics.

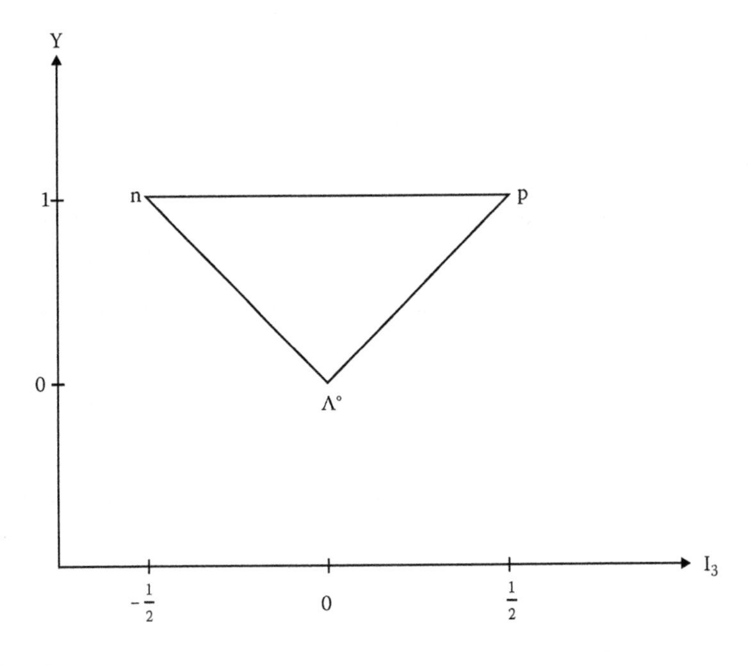

FIGURE 4.13
Baryon triplet for IR 3 of SU(3).

Similarly, the two particles associated with the weights **m**(2) and **m**(3) have, respectively,

$$\left(I_3 = -\frac{1}{2}, Y = 1, Q = 0\right)$$

$$\left(I_3 = 0, Y = 0, Q = 0\right)$$

The particle associated with the weight **m**(2) is a neutron, whereas the other particle is a lambda. The corresponding diagram with I_3 and Y as axes is shown in Figure 4.13. Thus, the irreducible representation 3, a multiplet of SU(3), can accommodate three particles—proton (p), neutron (n), and lambda (Λ°)—with ($I_3 = 1/2$, $Y = 1$, $Q = 1$), ($I_3 = -1/2$, $Y = 1$, $Q = 0$), ($I_3 = 0$, $Y = 0$, $Q = 0$), respectively. They constitute an isospin doublet (p,n) and an isospin singlet Λ°. Since these three particles p, n, Λ° are accommodated in the fundamental representation 3 of SU(3), these were considered as fundamental particles. More particles can be accommodated in other IRs of SU(3). Let us examine it.

The IR 8 of SU(3) has eight weights and can accommodate eight particles. The weights of this IR have already been determined:

$$\mathbf{m}(1)=\left(\frac{1}{\sqrt{3}},\ 0\right)$$
$$\mathbf{m}(2)=\left(\frac{1}{2\sqrt{3}},\ \frac{1}{2}\right)$$
$$\mathbf{m}(3)=\left(\frac{1}{2\sqrt{3}},\ -\frac{1}{2}\right)$$
$$\mathbf{m}(4)=\left(-\frac{1}{\sqrt{3}},\ 0\right)$$
$$\mathbf{m}(5)=\left(-\frac{1}{2\sqrt{3}},\ -\frac{1}{2}\right)$$
$$\mathbf{m}(6)=\left(-\frac{1}{2\sqrt{3}},\ \frac{1}{2}\right)$$
$$\mathbf{m}(7)=(0,\ 0)$$
$$\mathbf{m}(8)=(0,\ 0)$$

Proceeding on the same lines as for the IR 3, the I_3 and Y values of the particles associated with these weights are

$$\mathbf{m}(1):\left(I_3=1, Y=\frac{2}{3}\right)$$
$$\mathbf{m}(2):\left(I_3=\frac{1}{2}, Y=\frac{5}{3}\right)$$
$$\mathbf{m}(3):\left(I_3=\frac{1}{2}, Y=-\frac{1}{3}\right)$$
$$\mathbf{m}(4):\left(I_3=-1, Y=\frac{2}{3}\right)$$
$$\mathbf{m}(5):\left(I_3=-\frac{1}{2}, Y=-\frac{1}{3}\right)$$
$$\mathbf{m}(6):\left(I_3=-\frac{1}{2}, Y=\frac{5}{3}\right)$$
$$\mathbf{m}(7):\left(I_3=0, Y=\frac{2}{3}\right)$$
$$\mathbf{m}(8):\left(I_3=0, Y=\frac{2}{3}\right)$$

But as we associate these weights with the observables, we notice that charges on these particles should be (remember $Q = I_3 + Y/2$)

$$Q_1 = \frac{4}{3}, \quad Q_2 = \frac{4}{3}, \quad Q_3 = \frac{1}{3}, \quad Q_4 = \frac{2}{3}$$

$$Q_5 = -\frac{2}{3}, \quad Q_6 = \frac{1}{3}, \quad Q_7 = \frac{1}{3}, \quad Q_8 = \frac{1}{3}$$

Since particles with fractional charges were not known, the Sakata model was rejected.

Another model, now called the *quark model*, was proposed independently by Gell-Mann[5] and Zweig.[6] According to this model, each baryon is made up of three fundamental particles called *quarks* denoted by q_1, q_2, and q_3, whereas each meson is built from a *quark–antiquark* pair, $q\bar{q}$. They assumed the following relations for the quarks accommodated in the IR 3 of SU(3):

$$\sqrt{3}H_1 = I_3$$

$$2H_2 = Y$$

so that

$$I_3 = \sqrt{3}m_1$$

$$Y = 2m_2$$

The first relationship is the same as that in the Sakata model, but the second relationship is different. Let us now make some explicit calculations. Again, the three weights of the irreducible representation 3 of SU(3) are

$$\mathbf{m}(1) = \left(\frac{1}{2\sqrt{3}}, \frac{1}{6}\right)$$

$$\mathbf{m}(2) = \left(-\frac{1}{2\sqrt{3}}, \frac{1}{6}\right)$$

$$\mathbf{m}(3) = \left(0, -\frac{1}{3}\right)$$

Three particles associated with the weights have the following characteristics:

$$\mathbf{m}(1) = \left(I_3 = \frac{1}{2},\ Y = \frac{1}{3} \right)$$

$$\mathbf{m}(2) = \left(I_3 = -\frac{1}{2},\ Y = \frac{1}{3} \right)$$

$$\mathbf{m}(3) = \left(I_3 = 0, Y = -\ \frac{2}{3} \right)$$

The charges on these particles, as given by the formula $Q = I_3 + Y/2$, are

$$Q_1 = \frac{2}{3},\quad Q_2 = -\frac{1}{3},\quad Q_3 = -\frac{1}{3}$$

These particles, *fundamental particles in this model,* are quarks and carry fractional charges. To accommodate baryons, each one assumed to be made of three quarks, we consider the decomposition of the Kronecker product of three fundamental representations into a sum of IRs. As we have

$$3 \times 3 \times 3 = 1 + 8 + 8 + 10$$

we may determine eight baryons that can be accommodated in the IR 8 of SU(3). The eight weights of this representation, as already determined, are

$$\mathbf{m}(1) = \left(\frac{1}{\sqrt{3}},\ 0 \right)$$

$$\mathbf{m}(2) = \left(\frac{1}{2\sqrt{3}},\ \frac{1}{2} \right)$$

$$\mathbf{m}(3) = \left(\frac{1}{2\sqrt{3}},\ -\frac{1}{2} \right)$$

$$\mathbf{m}(4) = \left(-\frac{1}{\sqrt{3}},\ 0 \right)$$

$$\mathbf{m}(5) = \left(-\frac{1}{2\sqrt{3}},\ -\frac{1}{2} \right)$$

$$\mathbf{m}(6) = \left(-\frac{1}{2\sqrt{3}},\ \frac{1}{2} \right)$$

$$\mathbf{m}(7) = (0,\ 0)$$

$$\mathbf{m}(8) = (0,\ 0)$$

In the quark model, the characteristics of the particles in the IR 8 are

$$\mathbf{m}(1): (I_3 = 1, Y = 0)$$
$$\mathbf{m}(2): \left(I_3 = \frac{1}{2}, Y = 1\right)$$
$$\mathbf{m}(3): \left(I_3 = \frac{1}{2}, Y = -1\right)$$
$$\mathbf{m}(4): (I_3 = -1, Y = 0)$$
$$\mathbf{m}(5): \left(I_3 = -\frac{1}{2}, Y = -1\right)$$
$$\mathbf{m}(6): \left(I_3 = -\frac{1}{2}, Y = 1\right)$$
$$\mathbf{m}(7): (I_3 = 0, Y = 0)$$
$$\mathbf{m}(8): (I_3 = 0, Y = 0)$$

The charges on these particles are given by

$$Q_1 = 1, Q_2 = 1, Q_3 = 0, Q_4 = -1$$
$$Q_5 = -1, Q_6 = 0, Q_7 = 0, Q_8 = 0$$

The eight baryons with these values of I_3, Y, and Q are, respectively,

$$\Sigma^+, \ p, \ \Xi^0, \ \Sigma^-, \ \Xi^-, \ n, \ \Sigma^0, \ \Lambda$$

The corresponding diagram for the baryon octet is shown in Figure 4.14. All these baryons were known at the time the model was proposed.

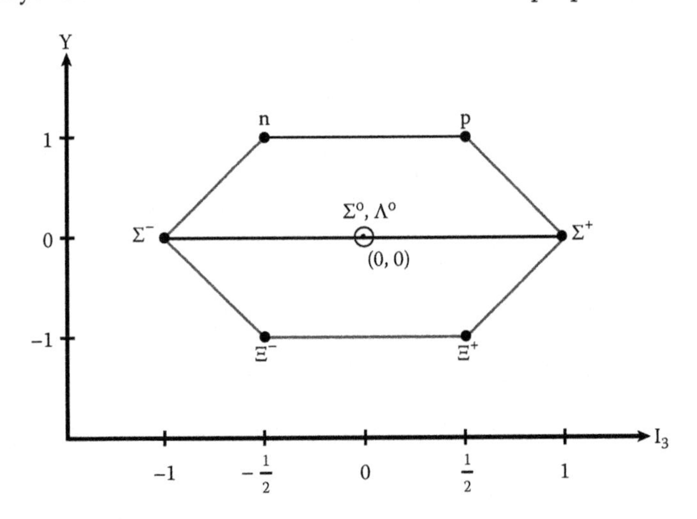

FIGURE 4.14
Baryon octet 8 with spin and parity $J^P = 1/2^+$.

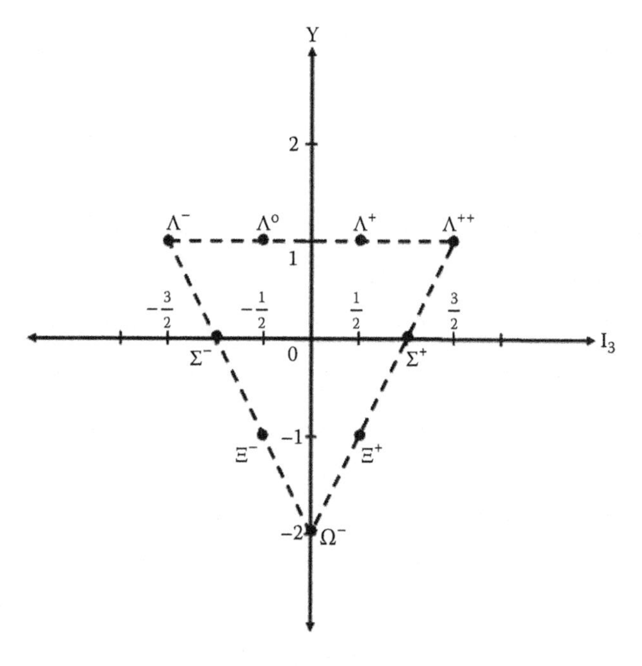

FIGURE 4.15
Baryon decuplet 10 with spin and parity $J^P = 3/2^+$.

The experimentally well-established baryon decuplet corresponding to the weights of IR $D^{(10)}(1,1) \equiv 10$ of SU(3) is shown in Figure 4.15. It consists of an isospin quartet Δ, an isospin triplet Σ^*, an isospin doublet Ξ^*, and an isospin singlet Ω^-. It is important to note that the Ω^- particle with characteristics specified by the baryon decuplet of SU(3) was predicted by the model and was discovered later on.

PROBLEM 4.5
Determine all 10 weights of the IR 10 of SU(3).

PROBLEM 4.6
Show that the mass splitting among the members of the baryon decuplet is given by

$$M_\Omega - M_{\Xi^*} = M_{\Xi^*} - M_{\Sigma^*} = M_{\Sigma^*} - M_\Lambda$$

PROBLEM 4.7
Derive the Gell-Mann–Okubo mass formula

$$M = a + bY + c\left[I(I+1) - \frac{1}{4}Y^2\right]$$

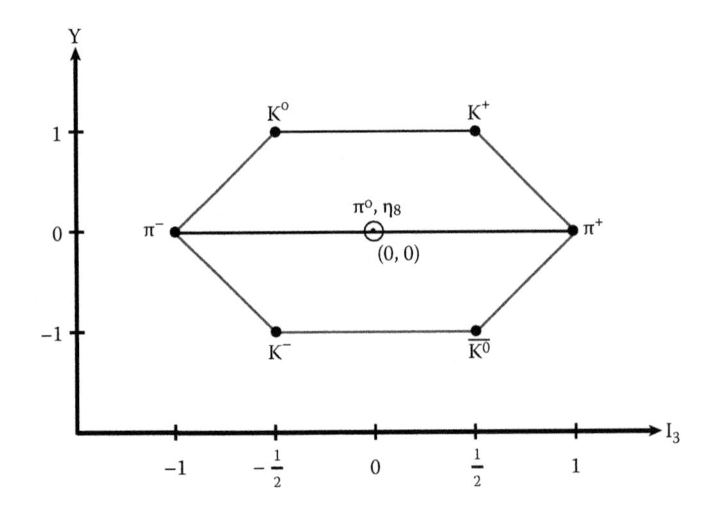

FIGURE 4.16
Meson octet.

$\bar{3}$ or 3* (≡ $D^{(3)}(0, 1)$) is another fundamental representation of SU(3). We know that

$$3 \times \bar{3} = 1 + 8$$

where $\bar{3}$ can accommodate three antiparticles. Eight mesons can be accommodated in the multiplet 8. The weights and I_3 and Y values are the same as for baryons. But the eight mesons with these values are

$$\pi^+, K^+, \overline{K^0}, \eta(548) \equiv \eta_8, \pi^\circ, \pi^-, K^-, \text{and } K^\circ$$

The corresponding diagram for eight mesons is given in Figure 4.16. It may be pointed out that throughout the discussion on continuous groups, our emphasis has been on SU(3). This is because, as was noticed by Ne'eman,[7] the components of the weights of irreducible representations of groups such as C_2 and G_2 could not be associated with the characteristics of strongly interacting particles. In fact, even for SU(3), the symmetry holds only approximately: it is a broken symmetry of strong interactions.

Let us next consider the application of spontaneous symmetry breaking in particle physics. We know that Lagrangian density serves as the basic object in the field theory. The demand that it should be symmetric with respect to a local gauge transformation makes it impossible to add the mass terms. The addition of a mass term or mass terms by hand destroys the local symmetry of the Lagrangian density. The theory then does not remain renormalizable, and its results are no more reliable. But there is another way of engendering masses of the gauge particles. The research carried out in solid-state physics shows that if the Lagrangian density of a system is symmetric under a certain local gauge transformation but its lowest energy state, called the ground state or the vacuum state, is not, then the gauge particles acquire masses. And, in spite of this, the theory remains renormalizable, and hence its results are reliable. If the Lagrangian density of a system is invariant under a gauge transformation but its vacuum state is not, then the SSB is said to occur or the system is said to have *hidden symmetry.*

Spontaneous symmetry breaking of the Lagrangian density $\mathcal{L}$ requires that, under a continuous group, $\mathcal{L}$ should be symmetric but the vacuum state should not be. However, the spontaneous symmetry breaking of $\mathcal{L}$ may not be exact. In this case, if a number of gauge fields are produced, the vacuum state of the system may be symmetric for some fields but not for others. This makes it possible to produce both massive and massless exchange particles.

Let us first discuss the spontaneous symmetry breaking of the Lagrangian density of a system under U(1) global gauge transformation. That is, let us find the constraints under which the Lagrangian density of a system possesses global symmetry with respect to the group U(1), but its ground state is asymmetric. With this objective in view, we investigate spontaneous symmetry breaking in the classical field theory, which involves an infinite number of degrees of freedom, by treating it purely classically but using the quantum mechanical language. Let us start with a complex scalar (spinless) field $\varphi(x)$ whose dynamics is described by the Lagrangian density $\mathcal{L}_1$:

$$\mathcal{L}_1 = (\partial_\mu \varphi)^* \,(\partial^\mu \varphi) - \{\mu^2 \varphi^* \varphi + \lambda(\varphi^* \varphi)^2\} \tag{4.4}$$

where μ^2 and λ are arbitrary real parameters. For convenience, we will be writing φ and $\varphi(x)$ interchangeably. The reason for choosing a quartic polynomial is that the corresponding quantum field theory is renormalizable; by including still higher-order $\varphi^*\varphi$ terms, the theory does not remain renormalizable. The first term on the right-hand side of Equation (4.4) corresponds to the kinetic energy, and the remaining terms in curly brackets constitute the potential energy V:

$$V(|\varphi|) = \mu^2 \varphi^* \varphi + \lambda(\varphi^* \varphi)^2 = \mu^2 \;|\varphi|^2 + \lambda \;|\varphi|^4$$

or

$$V(\rho) = \mu^2 \rho^2 + \lambda \rho^4 \tag{4.5}$$

where

$$\rho = |\varphi|$$

The Lagrangian density $\mathcal{L}_1$ is invariant under U(1) global gauge transformation

$$\varphi(x) \to e^{-i\theta} \varphi(x) = \wedge(\theta)\varphi(x)$$

as under this transformation the derivative $(\partial_\mu \varphi)$ transforms the same way as the field $\varphi(x)$ itself. But what about its ground state? We will show that it is not invariant under this very transformation provided $\lambda > 0$ and $\mu^2 < 0$: under these conditions, SSB occurs. To prove this, we notice that the Hamiltonian density of the system is given by

$$H = T + V = |\partial\varphi/\partial t|^2 + |\nabla\varphi|^2 + V(|\varphi|)$$

We are interested in finding the behavior of the vacuum state for which, by definition, the energy of the system is minimum. That is, we have to determine that value of the field φ, which gives the minimum field energy, that is, the lowest value of H. The first two terms of H are positive definite (i.e., these cannot be negative). Both the terms vanish for φ = constant, that is, when φ is independent of space as well as time coordinates. This means that the minimum value of the kinetic energy is zero and occurs for φ = constant. What is this constant equal to? This is determined by the requirement that it should give the minimum value of the remaining term in the expression for H—that is, for the potential energy $V(|\varphi|)$. This constant value of φ also gives the minimum value of total energy because for φ = constant, the value of the kinetic energy terms is minimum and equal to zero. To obtain this constant value of φ, we proceed as follows.

The potential energy $V(\rho = |\varphi|)$ given by

$$V(\rho) = \mu^2\rho^2 + \lambda\,\rho^4$$

where $\rho = |\varphi|$ can have a local minimum only if $\lambda > 0$, because otherwise $V(\rho) \to -\infty$, as $\rho \to \infty$. That is, if $\lambda < 0$, the potential is unbounded from below and has no state of minimum energy. For determining the value of ρ (= $|\varphi|$) and ultimately that of φ that gives the minimum value of the potential energy $V(\rho)$, we first write the first and second derivatives of the expression for the potential energy $V(\rho)$ with respect to ρ:

$$\frac{\partial V}{\partial \rho} = 2\mu^2\rho + 4\lambda\rho^3$$

and

$$\frac{\partial^2 V}{\partial \rho^2} = 2\mu^2 + 12\lambda\rho^2$$

The maxima and minima for the potential energy $V(\rho)$ with respect to ρ are obtained by putting $\partial V/\partial\rho = 0$. This gives

$$\rho\,(\mu^2 + 2\lambda\,\rho^2) = 0$$

That is, the maxima and minima of $V(\rho)$ occur either for $\rho = 0$ or for $\rho^2 = -\mu^2/2\lambda$, $\lambda > 0$. For any one of these values of ρ, the minimum value of potential energy will occur when the second derivative is positive.

For $\rho = 0$, the second derivative of $V(\rho)$ with respect to ρ, viz. $2\mu^2 + 12\lambda\rho^2$, gives $2\mu^2$. This is positive if μ^2 is positive. Therefore, for $\rho = 0$, the minimum value of $V(\rho)$ occurs for $\mu^2 > 0$. This minimum value of $V(\rho)$ is obtained by substituting $\rho = 0$ in Equation (4.5):

$$V_{min}\,(\rho = 0) = 0$$

This analysis shows that, for $\rho = 0$, the potential energy and consequently the total energy of the system has its minimum value equal to zero provided $\mu^2 > 0$. As $\rho^2 = |\varphi|^2$, where $\varphi = 1/\sqrt{2}\,(\varphi_1 + i\varphi_2)$ with $1/\sqrt{2}\ \varphi_1$ and $1/\sqrt{2}\ \varphi_2$ as real and imaginary parts of the field φ, this fact can also be expressed by stating that the potential energy, and consequently the total energy has its minimum value when $\rho^2 = |\varphi|^2 = 1/2(\varphi^2{}_1 + \varphi^2{}_2) = 0$, that is, only when $\varphi_1 = 0$ and $\varphi_2 = 0$ and hence when $\varphi = 0$. Since $\varphi_1 = 0$ and $\varphi_2 = 0$ are the only values of φ_1 and φ_2 that satisfy the constraint $\varphi^2{}_1 + \varphi^2{}_2 = 0$, the value zero of the field φ corresponding to the ground state is unique. Thus, under U(1) global gauge transformation—that is, for $\varphi\,(x) \rightarrow e^{-i\theta}\varphi\,(x)$—the ground state remains invariant. That is, both the Lagrangian density and the ground state of the system are symmetric with respect to this transformation; hence, spontaneous symmetry breaking does not occur. The potential energy graph as plotted against the field is a smooth bowl. The graph between φ and $V(\varphi)$ is shown in Figure 4.17.

Let us next consider the second value of ρ, viz. $\rho^2 = -\mu^2/2\lambda$, that can give a minimum value of the field energy. For this value of ρ^2, the second derivative

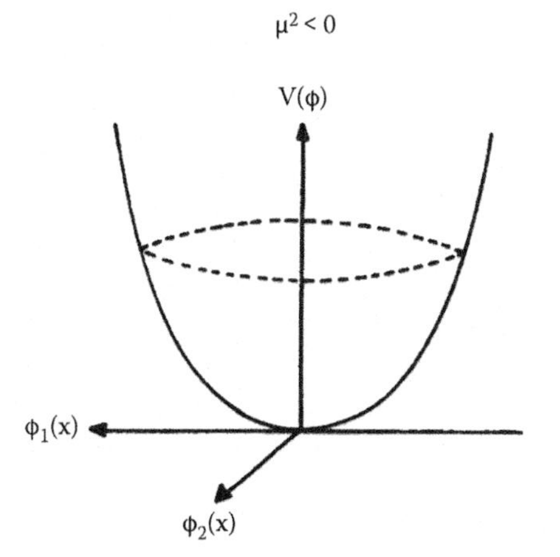

FIGURE 4.17
The potential function for positive μ^2.

of $V(\rho)$ is given by $\partial^2V/\partial\rho^2 = -4\mu^2$. This is positive for $\mu^2 < 0$. Hence, the potential energy $V(\rho)$ and consequently the field energy has a minimum at $\rho^2 = -\mu^2/2\lambda$ provided that μ^2 is negative.

The value of the potential energy at this point is

$$V_{min}\left(\rho^2 = -\frac{1}{2}\frac{\mu^2}{\lambda}\right) = -\frac{1}{4}\frac{\mu^4}{\lambda}$$

Since the complex scalar field φ is related to ρ by $\rho^2 = |\varphi|^2$, where $\varphi = 1/\sqrt{2}\ (\varphi_1 + i\varphi_2)$; the potential energy and consequently the total energy has its minimum value $-\mu^4/4\lambda$ and φ_1 and φ_2 satisfy the equation

$$\rho^2 = |\varphi|^2 \frac{(\varphi_1^2 + \varphi_2^2)}{2} = -\frac{1}{2}\frac{\mu^2}{\lambda}$$

for example,

$$\varphi_1^2 + \varphi_2^2 = -\frac{\mu^2}{\lambda} = v^2$$

This is the equation of a circle of radius v in $\varphi_1\varphi_2$-plane. This equation, a constraint, is satisfied by an infinite number of points, each point corresponding to a different pair of φ_1 and φ_2 values lying on the circumference of the circle. All the points lying on this circle in the $\varphi_1\varphi_2$-plane have different values of the field components $1/\sqrt{2}\ \varphi_1$ and $1/\sqrt{2}\ \varphi_2$ but possess the same value of the magnitude of the field φ and give the same minimum value of the field energy. Hence, there is a circle of degenerate minima in the complex φ-plane (i.e., in the $\varphi_1\varphi_2$-plane) with radius v given by $v = \sqrt{-\mu^2/\lambda}$, different field values for minimum field energy corresponding to an infinite number of points on the circle. In fact, the constraint expressed by equation $\rho^2 = |\varphi|^2 = -\mu^2/2\lambda$ does not fix the phase of φ defined as $\tan^{-1} \varphi_2/\varphi_1$: the magnitude of the field at every point on the circle is the same, but its direction changes. Hence, there exists an infinite number of pairs of φ_1 and φ_2 values, each pair satisfying the same equation $\varphi^2{}_1 + \varphi^2{}_2 = v^2$ and all giving the field φ for which the energy is minimum. The vacuum state is infinitely degenerate. Under U(1) global gauge transformation, $\varphi \rightarrow e^{-i\theta}\ \varphi$, the phase of a field changes. Thus, this transformation changes a certain vacuum state represented by a point on the circumference of the $\varphi_1\varphi_2$-circle into another vacuum state represented by another point on the circumference of the same circle so that the ground state is not invariant under U(1) global gauge transformation. In other words, under U(1), the ground states transform among themselves. As already stated, this way of obtaining an asymmetric ground state when the Lagrangian density itself is symmetric is known as spontaneous symmetry breaking. This asymmetry of the ground state is not due to the addition of a nonsymmetric term to the Lagrangian density but is a consequence of the arbitrary choice of one of the degenerate ground (vacuum) states.

Now every point on the circle represents a field value $v/\sqrt{2}$ for the ground state. Let an arbitrary real number δ be the phase of the field φ_0 for the minimum energy. Then the energy of the field will be minimum at values of φ_0 given by

$$\varphi_0 = \sqrt{-\mu^2/2\lambda}\, e^{i\delta} = v/\sqrt{2}\, e^{i\delta}$$

The vacuum expectation value (VEV) of the field is not zero but is given by

$$< \varphi > = \sqrt{-\mu^2/2\lambda} = v/\sqrt{2}$$

Moreover, as we have already seen, a local maximum occurs at $\rho = 0$. Thus, we have a local maximum of $V(|\varphi|)$ at the origin ($\varphi_1 = \varphi_2 = 0$) in the $\varphi_1\varphi_2$-plane. Then the potential starts decreasing and minimum of $V(|\varphi|)$ occurs at $\rho^2 = |\varphi|^2 = -\mu^2/2\lambda = v^2/2$. The graph between φ and $V(\varphi)$ is plotted in Figure 4.18. It looks like a Mexican hat!

To summarize, for a complex scalar field φ with $V(|\varphi|) = \mu^2(\varphi^*\varphi) + \lambda(\varphi^*\varphi)^2$, the spontaneous symmetry breaking occurs for $\lambda > 0$ and $\mu^2 < 0$. The Lagrangian density is symmetric with respect to U(1) global gauge transformations but not the infinitely degenerate ground state.

By just looking at the expression for the Lagrangian density $\mathcal{L}_1$ given in Equation (4.4), we cannot directly observe the particle spectrum produced by SSB. As we are not able to obtain the exact solution of the problem, perturbation theory has to be used. But, for $\mu^2 < 0$, the point $\varphi = 0$ where the local maximum occurs represents an unstable field position (unstable configuration), and therefore it is not possible to develop perturbation theory in terms

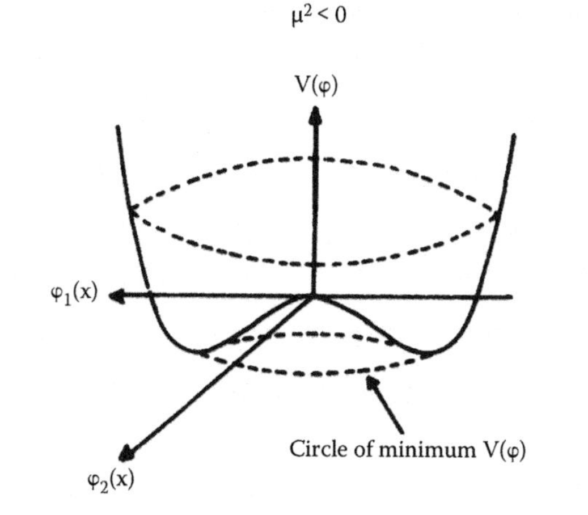

FIGURE 4.18
The potential function for negative μ^2.

of small departures from this point. As any point on the circle $\varphi^2_1 + \varphi^2_2 = v^2$, being a minimum of $V(|\varphi|)$, is a possible stable configuration, the perturbation theory can be developed about it, and that is what we are going to do. For this purpose, we translate the field $\varphi(x)$ to shift the origin to the minimum value of potential, that is, to any one point on this circle in the φ-plane. There is therefore no loss of generality by choosing a point on the circle lying on real axis in the complex φ-plane as the vacuum state. In other words, the direction of the vacuum, although arbitrary, is conventionally chosen along the real axis $\varphi_1(x)$ in the complex φ-plane. Since any point on the circle can be obtained from any given point on it by applying the gauge transformation $\varphi(x) \to \varphi'(x) = e^{-i\theta}\,\varphi(x)$, all the degenerate vacuum states represented by points on the circle are obtained by this transformation.

We now define real fields $\eta(x)$ and $\chi(x)$ such that

$$\eta(x) = \varphi_1(x) - v$$

and

$$\chi(x) = \varphi_2(x)$$

These equations yield

$$\varphi_1(x) = v + \eta(x)$$

and

$$\varphi_2(x) = \chi(x)$$

so that

$$\varphi(x) = \frac{1}{\sqrt{2}}[\varphi_1(x) + i\varphi_2(x)] = \frac{1}{\sqrt{2}}[v + \eta(x) + i\chi(x)]$$

Substituting this expression for $\varphi(x)$ in Equation (4.4) and making use of the relation $-\mu^2 = \lambda v^2$, we obtain

$$\begin{aligned}
\mathcal{L}_2 &= \frac{1}{2}\partial_\mu(v + \eta - i\chi)\partial^\mu(v + \eta + i\chi) \\
&\quad + \frac{1}{2}\lambda v^2(v + \eta - i\chi)(v + \eta + i\chi) \\
&\quad - \frac{\lambda}{4}[(v + \eta - i\chi)(v + \eta + i\chi)]^2 \\
&= \frac{1}{2}(\partial_\mu \eta)(\partial^\mu \eta) + \frac{1}{2}(\partial_\mu \chi)(\partial^\mu \chi) \\
&\quad - \frac{\lambda}{4}(\eta^2 - \chi^2)^2 + \frac{\lambda}{4}v^4 - \frac{1}{2}(2\lambda v^2)\eta^2 \\
&\quad - \lambda v\eta^3 - \lambda v\eta\chi^2
\end{aligned}$$

Let us consider $\mathcal{L}_2$ as a quantum theoretical Lagrangian density with η and χ as basic real fields. Then, by perturbation about the new origin corresponding to the minimum energy (vacuum state) and retaining terms up to second order, a comparison with the Klein–Gordon equation shows that η and χ are two real Klein–Gordon fields so that there are two particles:

One corresponds to the η-field, having standard kinetic energy term $1/2(\partial_\mu\eta)(\partial^\mu\eta)$ has a (mass)^2 $m^2_\eta = 2\lambda v^2$.

The other corresponds to the χ-field, having the standard kinetic energy term $1/2(\partial_\mu\chi)(\partial^\mu\chi)$ is massless: $m^2_\chi = 0$, because a term proportional to χ^2 is absent.

Both the particles are spinless. Thus, the spontaneous breaking of U(1) global gauge symmetry produces two spin-zero particles, one of which possesses a mass $\sqrt{2\lambda v^2}$ while the other is massless. The massless spin-zero particle is called a *Goldstone boson.* Since the Goldstone bosons do not occur in nature, this analysis is only of academic interest.

Let us next see what happens if instead of considering spontaneous breaking of global gauge symmetry, we consider the spontaneous breaking of *local* gauge symmetry. We will show that in this case the unwanted massless spin-zero boson disappears and is replaced by a massive spin-one boson. The number of massive bosons is equal to the generators of the symmetry group if the SSB is exact; otherwise, it is less than that. This is known as a *Higgs mechanism* and is described next in detail.

Consider again the Lagrangian density

$$\mathcal{L}_1 = (\partial_\mu\varphi)^*(\partial^\mu\varphi) - \mu^2\varphi^*\varphi - \lambda\,(\varphi^*\varphi)^2$$

This is invariant under U(1) global gauge transformations. The spontaneous symmetry breaking occurs for $\mu^2 < 0$ and $\lambda > 0$, but only spinless particles are produced. To obtain massive vector particles, we can construct the corresponding Lagrangian density that is invariant under U(1) local gauge transformations and for which SSB occurs for $\mu^2 < 0$. The modified Lagrangian density is

$$\mathcal{L}_{\text{local}} = (D_\mu\varphi)^*(D^\mu\varphi) - \mu^2\varphi^*\varphi - \lambda(\varphi^*\varphi)^2 - \frac{1}{4}F_{mv}F^{mv} \tag{4.6}$$

where D_μ and $F_{\mu\nu}$ are defined by

$$D_\mu = (\partial_\mu - iqA_\mu)$$

and

$$F_{\mu\nu} = \partial_\mu A_\nu - \partial_\nu A_\mu$$

This Lagrangian density is invariant under the combined transformations

$$\varphi \rightarrow e^{-i\theta(x)}\,\varphi$$

and

$$A_\mu \rightarrow A'_\mu = A_\mu - \frac{1}{q}\partial_\mu \theta(x)$$

An analysis similar to that carried out for U(1) global gauge symmetry shows that, for $\mu^2 < 0$ and $\lambda > 0$, there is a spontaneous breaking of local U(1) gauge symmetry: a ring of degenerate ground states exists. We must again shift the field to rewrite $\mathcal{L}_{\text{local}}$ in terms of displacement from the physical vacuum at $|\varphi| = v/\sqrt{2}$. Then

$$\varphi(x) = \frac{(v + \eta + i\chi)}{\sqrt{2}}$$

Substituting this expression in Equation (4.6), we get

$$\begin{aligned}\mathcal{L}_{\text{local}} = &-\frac{1}{4}F_{\mu\nu}F^{\mu\nu}\\ &+\left[(\partial_\mu + iqA_\mu)\frac{(v+\eta-i\chi)}{\sqrt{2}}\right.\\ &\left.(\partial_\mu + iqA_\mu)\frac{(v+\eta+i\chi)}{\sqrt{2}}\right]\\ &-\frac{\mu^2}{2}(v+\eta-i\chi)(v+\eta+i\chi)\\ &-\frac{\lambda}{4}[(v+\eta-i\chi)(v+\eta+i\chi)]\end{aligned}$$

This is the expression for $\mathcal{L}_{\text{local}}$ in terms of new real field variables η and χ.

Using the relation $-\mu^2 = \lambda v^2$ and simplifying, we get

$$\begin{aligned}\mathcal{L}_{\text{local}} = &-\frac{1}{4}F_{\mu\nu}F^{\mu\nu} + \frac{1}{2}q^2v^2A_\mu A^\mu\\ &+\frac{1}{2}(\partial_\mu\eta)(\partial^\mu\eta) + \frac{1}{2}(\partial_\mu\chi)(\partial^\mu\chi)\\ &-\frac{1}{2}(2\lambda v^2)\eta^2 - qvA_\mu(\partial_\mu\chi) + \frac{\lambda}{4}v^4\\ &+\text{cubic and higher-order terms}\end{aligned}$$

The terms $1/2\,(\partial_\mu\eta)(\partial^\mu\eta)$ and $-1/2(2\lambda\, v^2)\eta^2$ in the previous expression show that the real field η exists and its quantum is massive in nature, where the square of the mass is $2\lambda v^2$. However, owing to the mixing of the fields A_μ and χ through the term $-qv\, A_\mu(\partial_\mu\chi)$, a physical interpretation of the remaining spectrum is not evident. So the particle spectrum may be manifest from the expression for the Lagrangian density, it is imperative that, up to the quadratic terms, the Lagrangian density does not contain coupling terms such as $-qv\, A_\mu(\partial_\mu\chi)$. To achieve our goal, we choose the phase angle θ, called here as *gauge function*, as χ/v. Then we may write the transformation laws for φ and A_μ as

$$\varphi \to e^{-\frac{i\chi}{v}}\varphi$$

$$A_\mu \to A'_\mu = A_\mu - \frac{1}{q} v\,\partial_\mu \chi$$

Since $\varphi = \dfrac{(v+\eta+i\chi)}{\sqrt{2}}$, we have

$$\varphi \to e^{-\frac{i\chi}{v}}\varphi \approx \left(1 - \frac{i\chi}{v}\right)\frac{(v+\eta+i\chi)}{\sqrt{2}}$$

$$\approx \frac{(v+\eta)}{\sqrt{2}}$$

where, in the expansion for $e^{-i\chi/v}$ and for the product of last two terms, the quadratic and higher-order terms have been ignored. A gauge transformation

$$\varphi \to e^{-\frac{i\chi}{v}}\varphi$$

transforms a complex field φ into a real field

$$\varphi \to \frac{(v+\eta)}{\sqrt{2}}$$

Such a gauge is called a unitary gauge. It has the advantage that in this gauge the theory does not have any degrees of freedom with negative probability. Thus, we may finally write

$$\varphi \to \frac{(v+\eta)}{\sqrt{2}}$$

$$A_\mu \to A_\mu - \frac{1}{q} v\,\partial\mu\chi$$

Substituting these expressions for φ and A_μ in Equation (4.6), the Lagrangian density is transformed as detailed as follows. We have

$$(D_\mu\varphi)*(D^\mu\varphi) = (\partial_\mu + iq\, A_\mu)\varphi * (\partial^\mu - iq\, A^\mu)\varphi$$

$$\to \{\partial_\mu + iq\, A'_\mu\}\frac{(v+\eta)}{\sqrt{2}}$$

$$\{\partial^\mu - iq\, A'^\mu\}\frac{(v+\eta)}{\sqrt{2}}$$

$$= \frac{1}{2}\left[(\partial_\mu \eta)(\partial^\mu \eta) + q^2 A'_\mu A'^\mu (v+\eta)^2\right]$$

$$-\mu^2\varphi*\varphi \to \frac{1}{2}\lambda v^2(v+\eta)(v+\eta)$$

$$= \left(\frac{\lambda}{2}\right)v^2(v^2+\eta^2+2v\eta)$$

and

$$-\lambda(\varphi*\varphi)^2 \to -\frac{\lambda}{4}[(v+\eta)(v+\eta)]^2$$

$$= -\frac{\lambda}{4}[v^4+\eta^4+6v^2\eta^2+4v\eta(v^2+\eta^2)]$$

The last two expressions for transformation yield

$$(-\mu^2\varphi*\varphi - \lambda\,\varphi*\varphi) \to \frac{\lambda}{4}v^2 - \lambda v^2\eta^2 - \lambda v\eta^3 - \frac{\lambda}{4}\eta^6$$

We also notice that

$$F_{\mu\nu} = \partial_\mu A_\nu - \partial_\nu A_\mu \to \partial_\mu A'_\nu - \partial_\nu A'_\mu = F'_{\mu\nu}$$

Therefore, we must have

$$-\frac{1}{4}F_{\mu\nu}F^{\mu\nu} = -\frac{1}{4}F'_{\mu\nu}F'^{\mu\nu}$$

Hence, the Lagrangian density given by Equation (4.6) is transformed to

$$\mathcal{L}'_{\text{local}} = -\frac{1}{4}F'_{\mu\nu}F'^{\mu\nu} + (\partial_\mu\eta)(\partial^\mu\eta) + \frac{1}{2}q^2v^2A'_\mu A'^\mu$$

$$+\frac{1}{2}q^2A'_\mu A'^\mu\eta(2v+\eta) - \frac{1}{2}(2\lambda\, v^2)\eta^2 - \lambda v\eta^3$$

$$-\frac{\lambda}{4}v\eta^6 + \frac{\lambda}{4}v^4$$

The last term in the Lagrangian density $\mathcal{L}'_{local}$ is a constant and has no significance in our analysis. For small oscillations about the minimum, the effective Lagrangian density involves terms up to second order only. Thus, when the gauge field is taken as A'_μ, then up to second order there are no terms coupling different particles. It may be emphasized that the gauge field is A'_μ and not A_μ. The particle spectrum can now be directly read from the quadratic terms as follows:

A spinless η-meson with $(\text{mass})^2 = 2\lambda v^2 = -2\mu^2$. This is known as *Higgs boson* or *Higgs particle.*

A massive vector particle associated with the gauge field A'_μ with $(\text{mass})^2 = q^2v^2 = -q^2\mu^2/\lambda$.

There is no χ-field and therefore no particle corresponding to it.

We notice that in going from spontaneous breaking of global gauge symmetry to spontaneous breaking of local gauge symmetry, the massless spin-zero Goldstone boson disappears and is replaced by a massive spin-one boson with a mass qv as a quantum of the associated gauge field A'_μ. This is expressed in a colorful language by stating that the massless vector gauge boson has eaten the massless Goldstone boson to become a massive vector boson! Moreover, a massive spin-zero particle, called Higgs particle, has been engendered.

The generalization with an r-parameter continuous symmetry group G will produce r massive gauge particles with spin 1. Such particles exist in nature, and the generalization, at least in principle, is consistent with experiment. Thus, SSB of local gauge symmetry under a continuous symmetry group and with a unitary gauge produces massive gauge particles.

We would like to give brief remarks about renormalization. A renormalizable theory gets rid of infinities that are introduced during computation and gives a one-to-one correspondence between theoretical and experimental values. The results of a theory are reliable to all orders of the coupling constant and at all energies only if it is renormalizable. 't Hooft proved that gauge theories even with massive intermediate vector bosons are renormalizable.[8] The significance of gauge theories stems from his proof as it makes their results reliable. He solved this problem in 1971 when he was a young Ph.D. scholar. In 1999, almost three decades after this significant contribution, 't Hooft shared the Nobel Prize in physics for the same. Pickering made a beautiful remark[9]: To the high energy physics community at large, 't Hooft was an unknown Dutch graduate student: not the most likely person to have solved a problem that had defeated some of the world's leading field theorists over a period of nearly two decades.

References

1. Kronig, R. and Weisskopf, V. F., *Collected Papers of Wolfgang Pauli*, vol. 1, Wiley-Interscience, New York, 1964.
2. Perelomov, A. M. and Popov, V.S. *Math. USSR Izv* 2, 1313, 1968.
3. Campoamor-Stursberg, R. *Monografías de la Real Academia de Ciencias de Zaragoza*, 37, 11, 2011.
4. Sakata, S. *Prog. Tehor. Phys.*, 16, 686, 1956.
5. Gell-Mann, *Physics Letters*, 8, 214, 1964.
6. Zweig, CERN reprint, 8409/Th. 412, 1964.
7. Ne'eman, Y., *Nucl. Phys.*, 26, 222, 1961.
8. t' Hooft, G., *Nucl. Phys.*, B 35, 167, 1971.
9. Pickering, A., *Constructing Quarks*, University of Chicago Press, Chicago, IL, p. 179, 1984.

Appendix A: Commutation Relations between the Generators of a Semisimple Lie Group

Consider an r-parameter semisimple Lie group of rank ℓ. It has a set of ℓ mutually commuting generators H_i, $i = 1, 2, \ldots, \ell$, the set of remaining $r - \ell$ generators being denoted by E_α, $\alpha = 1, 2, \ldots, r - \ell$. Let L_A stand for any one of the generators of the group. Let us choose the parameterization such that all the generators L_A, $A = 1, 2, \ldots, r$ are Hermitian. We know that the commuting generators H_i or the corresponding matrices, say C_i, can always be diagonalized simultaneously. Let us now determine the commutation relations between the generators L_A.

Let us first consider the commutator of H_i with H_j. Since the commutator of any two generators of a group is a linear combination of the generators of the same group, we have

$$[H_i, H_j] = C_{ij}^{D} L_D$$

where there is a summation over D. But as H-type generators mutually commute, we must have

$$[H_i, H_j] = 0 \tag{A.1}$$

Comparing these two relations, we notice that

$$C_{ij}^{D} L_D = 0$$

Since the generators of a group are linearly independent, the coefficient of each L_D must be zero. That is, we must have

$$C_{ij}^{D} = 0 \text{ for all D}$$

Next we consider the commutator of H_i with E_α. We have

$$[H_i, E_\alpha] = C_{i\alpha}^{D} L_D$$

where there is a summation over D. But for $D \neq \alpha$, the nondiagonal elements $C_{i\alpha}^{D}$ of the diagonal matrix C_i must vanish

$$C_{i\alpha}^{D} = 0, \text{ for } D \neq \alpha$$

Therefore, this commutation relation reduces to

$$[H_i, E_\alpha] = C^{\alpha}_{i\alpha}E_\alpha = r_i(\alpha)E_\alpha \tag{A.2}$$

where $r_i(\alpha) = C^{\alpha}_{i\alpha}$ (no summation over α).

Let us next find the commutator of E_α and E_β. This is in general given by

$$[E_\alpha, E_\beta] = C^{D}_{\alpha\beta}L_D = C^{i}{}_{\alpha\beta}H_i + C^{\gamma}{}_{\alpha\beta}E_\gamma \tag{A.3}$$

Let us simplify it. We know that any three generators L_A, L_B, and L_D satisfy Jacobi's identity:

$$[L_A, [L_B, L_D]] + [L_B, [L_D, L_A]] + [L_D, [L_A, L_B]] = 0$$

We can write this as

$$C^{E}{}_{BD}[L_A, L_E] + C^{E}{}_{DA}[L_B, L_E] + C^{E}{}_{AB}[L_D, L_E] = 0$$

or

$$C^{E}{}_{BD}C^{F}{}_{AE}L_F + C^{E}{}_{DA}C^{F}{}_{BE}L_F + C^{E}{}_{AB}C^{F}{}_{DE}L_F = 0$$

or

$$(C^{E}{}_{BD}C^{F}{}_{AE} + C^{E}{}_{DA}C^{F}{}_{BE} + C^{E}{}_{AB}C^{F}{}_{DE})L_F = 0$$

Since the generators of a group are linearly independent, the coefficient of each generator must vanish:

$$C^{E}{}_{BD}C^{F}{}_{AE} + C^{E}{}_{DA}C^{F}{}_{BE} + C^{E}{}_{AB}C^{F}{}_{DE} = 0$$

For the three generators $H_i \equiv L_A$, $E_\alpha \equiv L_B$, and $E_\beta \equiv L_D$, the previous equation becomes

$$C^{E}{}_{\alpha\beta}C^{F}{}_{iE} + C^{E}{}_{\beta i}C^{F}{}_{\alpha E} + C^{E}{}_{i\alpha}C^{F}{}_{\beta E} = 0$$

However, the structure constants are antisymmetric with respect to their lower indices: $C^{D}{}_{AB} = -C^{D}{}_{BA}$. Using this property, we can write this equation as

$$C^{E}{}_{\alpha\beta}C^{F}{}_{iE} - C^{E}{}_{i\beta}C^{F}{}_{\alpha E} + C^{E}{}_{i\alpha}C^{F}{}_{\beta E} = 0$$

or

$$C^{F}{}_{\alpha\beta}r_i(F) - C^{F}{}_{\alpha\beta}r_i(\beta) + C^{F}{}_{\beta\alpha}r_i(\alpha) = 0$$

or

$$C^F_{\alpha\beta}(r_i(F) - r_i(\beta) - r_i(\alpha)) = 0 \quad \text{(A.4)}$$

where $r_i(F)$ stands for C^F_{iF} (no summation over F) = $(C_i)^F_F$, and the FF shows the diagonal element of C_i.

If F = j, then $r_i(F) = r_i(j) \equiv C^j_{ij} = 0$ by virtue of Equations (A.1), and consequently Equation (A.4) reduces to

$$C^j_{\alpha\beta}[r_i(\alpha) + r_i(\beta)] = 0$$

This leads to

$$C^j_{\alpha\beta}(\alpha + \beta) = 0 \quad \text{(A.5)}$$

where α and β are two nonzero roots of the group.

For F = γ, Equation (A.4) yields

$$C^\gamma_{\alpha\beta}[r_i(\gamma) - r_i(\beta) - r_i(\alpha)] = 0$$

This leads to

$$C^\gamma_{\alpha\beta}(\gamma - \beta - \alpha) = 0 \quad \text{(A.6)}$$

where α, β, and γ are three nonvanishing roots of the group.

Now, if $\alpha + \beta = 0$, that is, if $\beta = -\alpha$, then $\alpha + \beta \neq \gamma$ because otherwise γ will be a zero root, which is not true. Consequently, by virtue of Equation (A.6), we have $C^\gamma_{\alpha\beta} = 0$. Therefore, Equation (A.3) reduces to

$$[E_\alpha, E_\beta] = C^i_{\alpha\beta}H_i$$

or

$$[E_\alpha, E_{-\alpha}] = r^i(\alpha)H_i \quad \text{(A.7)}$$

where $r^i(\alpha) = C^i_{\alpha\beta} = C^i_{\alpha,-\alpha}$, because $\beta = -\alpha$.

If $\alpha + \beta \neq 0$, then by virtue of Equation (A.5), the structure constant $C^i_{\alpha\beta} = 0$ and using Equation (A.3), we have

$$[E_\alpha, E_\beta] = C^\gamma_{\alpha\beta}\, E_\gamma \quad \text{(A.8)}$$

Now two possibilities arise:

(i) If $\alpha + \beta = \gamma$, then by Equation (A.5), $C^\gamma_{\alpha\beta} = N_{\alpha\beta}$, say, and writing $E_{\alpha+\beta}$ for E_γ, Equation (A.8) yields

$$[E_\alpha, E_\beta] = N_{\alpha\beta}\, E_{\alpha+\beta} \quad \text{(A.9a)}$$

(ii) If $\boldsymbol{\alpha} + \boldsymbol{\beta} \neq \boldsymbol{\gamma}$, then by Equation (A.5), $C^{\gamma}_{\alpha\beta} = 0$. Consequently, Equation (A.8) yields

$$[E_\alpha, E_\beta] = 0 \tag{A.9b}$$

Hence

$$\begin{aligned} [E_\alpha, E_\beta] &= N_{\alpha\beta}\, E_{\alpha+\beta}, \text{ if } \boldsymbol{\alpha} + \boldsymbol{\beta} \text{ is a (nonzero) root} \\ &= 0, \text{ otherwise} \end{aligned}$$

To sum up, the generators of a semisimple Lie group satisfy the following commutation relations:

$$\begin{aligned} &[H_i, H_j] = 0 \\ &[H_i, E_\alpha] = r_i(\alpha) E_\alpha \\ &[E_\alpha, E_{-\alpha}] = r^i(\alpha) H_i \\ &[E_\alpha, E_\beta] = N_{\alpha\beta} E_{\alpha+\beta}, \text{ if } \boldsymbol{\alpha} + \boldsymbol{\beta} \text{ is a (nonzero) root} \\ &\qquad\quad\; = 0, \text{ otherwise} \end{aligned}$$

where, with a suitable normalization, we can have

$$r_i(\alpha) = r^i(\alpha)$$

and

$$\sum_{\alpha=1}^{r-\ell} r_i(\alpha)\, r_j(\alpha) = \delta_{ij}$$

Appendix B

Theorem B.1

If $\boldsymbol{\alpha}$ and $\boldsymbol{\beta}$ are two nonvanishing roots of a semisimple Lie group, then $g(\alpha/\beta) = 2\frac{\alpha \cdot \beta}{\beta\ \beta}$ is an integer, and $\boldsymbol{\alpha} - g(\alpha/\beta)\boldsymbol{\beta}$ is also a root.

PROOF

Since $\boldsymbol{\beta}$ is a root, $-\boldsymbol{\beta}$ should also be a root of the group. Let E_α, E_β, and $E_{-\beta}$ be the generators corresponding to the roots $\boldsymbol{\alpha}$, $\boldsymbol{\beta}$, and $-\boldsymbol{\beta}$. Then, by using the Jacobi identity, we have

$$[E_\alpha, [E_\beta, E_{-\beta}]] + [E_\beta, [E_{-\beta}, E_\alpha]] + [E_{-\beta}, [E_\alpha, E_\beta]] = 0$$

or

$$[E_\alpha, r_i(\beta)H_i] + [E_\beta, N_{-\beta\alpha}, E_{-\beta+\alpha}] + [E_{-\beta}, N_{\alpha\beta}, E_{\alpha+\beta}] = 0$$

or

$$r_i(\beta)[E_\alpha, H_i] + N_{-\beta\alpha}\,[E_\beta, E_{-\beta+\alpha}] + N_{\alpha\beta}[E_{-\beta}, E_{\alpha+\beta}] = 0$$

Notice that $N_{\beta\alpha}$ and $N_{\alpha\beta}$ will be different from zero only if $\boldsymbol{\alpha} \pm \boldsymbol{\beta}$ are nonvanishing roots. The previous equation yields

$$r_i(\beta)\{-r_i(\alpha)E_\alpha\} + N_{-\beta\alpha}\,N_{\beta,-\beta+\alpha}E_\alpha + N_{\alpha\beta}N_{-\beta,\alpha+\beta}E_\alpha = 0$$

or

$$(-\boldsymbol{\alpha} \cdot \boldsymbol{\beta} + N_{-\beta\alpha}N_{\beta,-\beta+\alpha} + N_{\alpha\beta}N_{-\beta,\alpha+\beta})E_\alpha = 0$$

Since the generator E_α cannot be equal to zero, we must have

$$\boldsymbol{\alpha} \cdot \boldsymbol{\beta} = N_{-\beta\alpha}N_{\beta,-\beta+\alpha} + N_{\alpha\beta}N_{-\beta,\alpha+\beta}$$

Replacing $\boldsymbol{\alpha}$ by $\boldsymbol{\alpha} + n\boldsymbol{\beta}$, where n is an integer, we get

$$\begin{aligned}(\boldsymbol{\alpha} + n\boldsymbol{\beta}) \cdot \boldsymbol{\beta} &= N_{-\beta,\alpha+n\beta}N_{\beta,-\beta+\alpha+n\beta} + N_{\alpha+n\beta,\beta}N_{-\beta,\alpha+n\beta+\beta} \\ \text{or} \quad (\boldsymbol{\alpha} + n\boldsymbol{\beta}) \cdot \boldsymbol{\beta} &= N_{-\beta,\alpha+n\beta}N_{\beta,\alpha+(n-1)\beta} + N_{\alpha+n\beta,\beta}N_{-\beta,\alpha+(n+1)\beta}\end{aligned} \tag{B.1}$$

Let us define μ_n by

$$\mu_n = N_{\alpha+n\beta,\beta} N_{-\beta,\alpha+(n+1)\beta} \tag{B.2a}$$

Then

$$\mu_{n-1} = N_{\alpha+(n-1)\beta,\beta} N_{-\beta,\alpha+n\beta} \tag{B.2b}$$

Subtracting Equation (B.2b) from Equation (B.2a), we get

$$\mu_n - \mu_{n-1} = N_{-\beta,\alpha+n\beta} N_{\beta,\alpha+(n-1)\beta} + N_{\alpha+n\beta,\beta} N_{-\beta,\alpha+(n+1)\beta}$$

where we have made use of the property: $N_{AB} = -N_{BA}$. By using Equation (B.1), this reduces to

$$\mu_n - \mu_{n-1} = (\boldsymbol{\alpha} + n\boldsymbol{\beta}) \cdot \boldsymbol{\beta}$$

or

$$\mu_n = \mu_{n-1} + (\boldsymbol{\alpha} + n\boldsymbol{\beta}) \cdot \boldsymbol{\beta} \tag{B.3}$$

Let us next consider the following two strings of commutators of E-type generators:

String 1:

$$[E_\alpha, E_\beta] = N_{\alpha\beta} E_{\alpha+\beta}$$
$$[E_{\alpha+\beta}, E_\beta] = N_{\alpha+\beta,\beta} E_{\alpha+2\beta}$$
$$[E_{\alpha+2\beta}, E_\beta] = N_{\alpha+2\beta,\beta} E_{\alpha+3\beta}$$
$$\cdots\cdots\cdots\cdots\cdots\cdots$$

String 2:

$$[E_\alpha, E_{-\beta}] = N_{\alpha,-\beta} E_{\alpha-\beta}$$
$$[E_{\alpha-\beta}, E_{-\beta}] = N_{\alpha-\beta,-\beta} E_{\alpha-2\beta}$$
$$[E_{\alpha-2\beta}, E_{-\beta}] = N_{\alpha-2\beta,-\beta} E_{\alpha-3\beta}$$
$$\cdots\cdots\cdots\cdots\cdots\cdots$$

In the first string, the structure constants will be different from zero only if

$$\boldsymbol{\alpha} + \boldsymbol{\beta}, \boldsymbol{\alpha} + 2\boldsymbol{\beta}, \boldsymbol{\alpha} + 3\boldsymbol{\beta} + \cdots$$

are nonvanishing roots. Since the number of roots of a semisimple Lie group is finite, the series of roots must terminate somewhere. Therefore, the string must terminate after a finite number of steps; that is, after a finite number of steps the commutator of two E's must be zero. The same is true for the second string. Suppose that the two strings terminate after j and k steps, respectively, so that j and k are nonnegative integers. Then the last equations of the strings with nonzero structure constants will be

$$[E_{\alpha+(j-1)\beta}, E_{\beta}] = N_{\alpha+(j-1)\beta,\beta}E_{\alpha+j\beta}$$

and

$$[E_{\alpha-(k-1)\beta}, E_{-\beta}] = N_{\alpha-(k-1)\beta,\beta}E_{\alpha-k\beta}$$

Since, after k steps, the commutator of two E's must be zero, we have

$$[E_{\alpha-k\beta}, E_{-\beta}] = N_{\alpha-k\beta,-\beta}E_{\alpha-(k+1)\beta} = 0$$

This yields

$$N_{\alpha-k\beta,-\beta} = 0 \tag{B.4a}$$

Similarly, since

$$[E_{\alpha+j\beta}, E_{\beta}] = N_{\alpha+j\beta,\beta}E_{\alpha+(j+1)\beta} = 0$$

we have

$$N_{\alpha+j\beta,\beta} = 0 \tag{B.4b}$$

Writing n = −k − 1, Equation (B.2a) yields

$$\mu_{-k-1} = N_{\alpha-(k+1)\beta,\beta}N_{-\beta,\alpha-k\beta}$$

or

$$\mu_{-k-1} = -N_{\alpha-(k+1)\beta,\beta}N_{\alpha-k\beta,-\beta} \tag{B.5}$$

Equations (B.5) and (B.4a) yield

$$\mu_{-k-1} = 0 \tag{B.6a}$$

From Equations (B.2a) and (B.4b), writing n = j, we get

$$\mu_j = N_{\alpha+j\beta,\beta}N_{-\beta,\alpha+(j+1)\beta} = 0 \tag{B.6b}$$

Now summing over n, Relation (B.3) yields

$$\sum_{n=-k}^{j}\mu_n = \sum_{n=-k}^{j}\mu_{n-1} + \sum_{n=-k}^{j}(\boldsymbol{\alpha}+n\boldsymbol{\beta})\cdot\boldsymbol{\beta}$$

or

$$\mu_{-k} + \mu_{-k+1} + \cdots + \mu_{j-1} + \mu_j = \mu_{-k-1} + \mu_{-k} + \cdots + \mu_{j-1}$$
$$+\sum_{n=-k}^{j} \alpha \cdot \boldsymbol{\beta} + \sum_{n=-k}^{j} n\boldsymbol{\beta} \cdot \boldsymbol{\beta}$$

or

$$\mu_j = \mu_{-k-1} + (k + j + 1)\, \boldsymbol{\alpha} \cdot \boldsymbol{\beta} + \underbrace{\{-k + (-k + 1) + \cdots + j\}}_{(k + j + 1) \text{ terms}}\, \boldsymbol{\beta} \cdot \boldsymbol{\beta}$$

or

$$\mu_j = \mu_{-k-1} + (k + j + 1)\, \boldsymbol{\alpha} \cdot \boldsymbol{\beta} + \frac{(k+j+1)\,(j-k)}{2}\, \boldsymbol{\beta} \cdot \boldsymbol{\beta}$$

Since, by virtue of Equations (B.6a) and (B.6b), $\mu_j = 0 = \mu_{-k-1}$, the previous equation reduces to

$$\boldsymbol{\alpha} \cdot \boldsymbol{\beta} + (j - k)\frac{1}{2}\boldsymbol{\beta} \cdot \boldsymbol{\beta} = 0$$

But as β is a nonvanishing root and the components of a root are always real, $\boldsymbol{\beta} \cdot \boldsymbol{\beta} > 0$. Therefore, we can divide this relation by $\boldsymbol{\beta} \cdot \boldsymbol{\beta}$. This yields, for example,

$$2\frac{\boldsymbol{\alpha} \cdot \boldsymbol{\beta}}{\boldsymbol{\beta} \cdot \boldsymbol{\beta}} = k - j$$

or

$$2\frac{\boldsymbol{\alpha} \cdot \boldsymbol{\beta}}{\boldsymbol{\beta} \cdot \boldsymbol{\beta}} = k - j = \text{an integer} = g\left(\frac{\alpha}{\beta}\right), \text{ say} \tag{B.7}$$

This proves the first part of the theorem.

To prove the second part, we recall that $g = k - j$ and k and j are nonnegative integers. Thus, $g \leq k$ and also $g \geq -j$, that is, $-j \leq g \leq k$. Now we know that $\boldsymbol{\alpha}, \boldsymbol{\alpha} - \boldsymbol{\beta}, \boldsymbol{\alpha} - 2\boldsymbol{\beta}, \ldots, \boldsymbol{\alpha} - k\boldsymbol{\beta}$, and $\boldsymbol{\alpha}, \boldsymbol{\alpha} + \boldsymbol{\beta}, \boldsymbol{\alpha} + 2\boldsymbol{\beta}, \ldots, \boldsymbol{\alpha} + j\boldsymbol{\beta}$ are roots. Thus, $\boldsymbol{\alpha} - n\boldsymbol{\beta}$ is a root if $-j \leq n \leq k$. Since g satisfies this condition, we conclude that $\boldsymbol{\alpha} - g\boldsymbol{\beta}$ is also a root. ■

Appendix C: Computation of Structure Constants

Let us determine the symmetry relations among the structure constants and calculate the value of $N_{\alpha\beta}$.

Suppose that $\boldsymbol{\alpha} + \boldsymbol{\beta}$ is a nonvanishing root of a semisimple Lie group. Let E_α, E_β, and $E_{-\alpha-\beta}$ be the generators corresponding to the roots $\boldsymbol{\alpha}$, $\boldsymbol{\beta}$, and $-\boldsymbol{\alpha}-\boldsymbol{\beta}$ of the group. Then by virtue of the Jacobi identity we have

$$[E_\alpha, [E_\beta, E_{-\alpha-\beta}]] + [E_\beta, [E_{-\alpha-\beta}, E_\alpha]] + [E_{-\alpha-\beta}, [E_\alpha, E_\beta]] = 0$$

or

$$[E_\alpha, N_{\beta,-\alpha-\beta}E_{-\alpha}] + [E_\beta, N_{-\alpha-\beta,\alpha}E_{-\beta}] + [E_{-\alpha-\beta}, N_{\alpha\beta}E_{\alpha+\beta}] = 0$$

or

$$N_{\beta,-\alpha-\beta}[E_\alpha, E_{-\alpha}] + N_{-\alpha-\beta,\alpha}[E_\beta, E_{-\beta}] + N_{\alpha\beta}[E_{-\alpha-\beta}, E_{\alpha+\beta}] = 0$$

or

$$N_{\beta,-\alpha-\beta}\, r_i(\alpha)H_i + N_{-\alpha-\beta,\alpha}\, r_i(\beta)\, H_i - N_{\alpha\beta}r_i(\alpha+\beta)H_i = 0$$

or

$$(N_{\beta,-\alpha-\beta}\, r_i(\alpha) + N_{-\alpha-\beta,\alpha}\, r_i(\beta) - N_{\alpha\beta}\, r_i(\alpha+\beta))\, H_i = 0$$

This relation will hold only if the coefficient of H_i is zero, that is,

$$N_{\beta-\alpha-\beta}\, r_i(\alpha) + N_{-\alpha-\beta,\alpha}\, r_i(\beta) - N_{\alpha\beta}r_i(\alpha+\beta) = 0$$

or

$$N_{\beta,-\alpha-\beta}\, r_i(\alpha) + N_{-\alpha-\beta,\alpha}\, r_i(\beta) = N_{\alpha\beta}r_i\,(\alpha+\beta)$$

However, as $r_i(\alpha) + r_i(\beta) = r_i(\alpha+\beta)$ for all i, the previous relation will hold for all i only if

$$N_{\alpha\beta} = N_{\beta,-\alpha-\beta} = N_{-\alpha-\beta,\alpha} \tag{C.1}$$

Because of the symmetry $\boldsymbol{\alpha} \leftrightarrow -\boldsymbol{\alpha}$ in the set of roots, it is always possible to choose, for the operators E_α, a normalization so that

$$N_{\alpha\beta} = -N_{-\alpha-\beta} \tag{C.2}$$

From Equations (C.1) and (C.2), we have

$$N_{\alpha\beta} = -N_{-\alpha,-\beta} = N_{\beta,-\alpha-\beta} = N_{-\alpha-\beta,\alpha} \tag{C.3}$$

These are the symmetry relations for structure constants.

To determine the value of $N_{\alpha\beta}$, we consider Equation (B.2a), namely,

$$\mu_n = N_{\alpha+n\beta,\beta}N_{-\beta,\alpha+(n+1)\beta} \tag{C.3a}$$

For n = 0, this reduces to

$$\mu_0 = N_{\alpha\beta}N_{-\beta,\alpha+\beta} \tag{C.4}$$

Now Equation (B.3) is

$$\mu_n = \mu_{n-1} + (\boldsymbol{\alpha} + n\,\boldsymbol{\beta}) \cdot \boldsymbol{\beta} \tag{B.3'}$$

Summing over n, we write it as

$$\mu_{-k} + \mu_{-k+1} + \cdots + \mu_{-1} + \mu_0$$

$$= \mu_{-k-1} + \mu_{-k} + \cdots + \mu_{-1} + (k+1)\ \boldsymbol{\alpha} \cdot \boldsymbol{\beta} - \frac{1}{2}(k+1)k\ \boldsymbol{\beta} \cdot \boldsymbol{\beta}$$

or

$$\mu_0 = (k+1)\,\boldsymbol{\alpha} \cdot \boldsymbol{\beta} - \frac{1}{2}(k+1)k\,\boldsymbol{\beta} \cdot \boldsymbol{\beta} \tag{B.6a}$$

or

$$\mu_0 = (k+1)\,\frac{1}{2}\,(k-j)\,\boldsymbol{\beta} \cdot \boldsymbol{\beta} - \frac{1}{2}(k+1)k\boldsymbol{\beta} \cdot \boldsymbol{\beta} \tag{B.7}$$

or

$$\mu_0 = -\frac{1}{2}\,j(k+1)\,\boldsymbol{\beta} \cdot \boldsymbol{\beta} \tag{C.5}$$

From Relations (C.4) and (C.5), we obtain

$$N_{\alpha\beta}N_{-\beta,\alpha+\beta} = -\frac{1}{2}j\,(k+1)\,\boldsymbol{\beta}\cdot\boldsymbol{\beta} \tag{C.6}$$

By virtue of Equation (C.3), this reduces to

$$N_{\alpha\beta} = \pm\sqrt{\frac{1}{2}j(k+1)\boldsymbol{\beta}\cdot\boldsymbol{\beta}}$$

This relation shows that the values of the structure constants are undetermined up to a sign, but they must be consistent in accordance with Relations (C.3). However, different consistent signs yield isomorphic and hence not distinct Lie groups.

Appendix D

Theorem D.1

If $\boldsymbol{\alpha}$ is a root of a semisimple Lie group and $\mathbf{m}$ is a weight of a representation of the group, then $g(m/\alpha) = 2\,\mathbf{m} \cdot \boldsymbol{\alpha}/(\boldsymbol{\alpha} \cdot \boldsymbol{\alpha})$ is an integer, and $\mathbf{m} - g(m/\boldsymbol{\alpha})\,\alpha$ is also a weight with the same multiplicity as $\mathbf{m}$.

PROOF

Consider an irreducible representation of a semisimple Lie group of rank ℓ. Let $\mathbf{H}$ represent the set of ℓ mutually commuting generators H_i, $i = 1, 2, \ldots, \ell$ of the group. Let $|m\rangle$ be an eigenstate of the generator $\mathbf{H}$, and let $\mathbf{m}$ be the corresponding weight. Then

$$\mathbf{H}\,|m\rangle = \mathbf{m}|m\rangle$$

Let $\boldsymbol{\alpha}$ be a root, the corresponding generator being denoted by E_α. Then

$$[\mathbf{H}, E_\alpha] = \boldsymbol{\alpha} E_\alpha$$

Operating it on the state $|m\rangle$, we get

$$\mathbf{H}\,E_\alpha\,|m\rangle - E_\alpha\,\mathbf{H}|m\rangle = \boldsymbol{\alpha}\,E_\alpha\,|m\rangle$$

or

$$\begin{aligned} \mathbf{H}\,E_\alpha\,|m\rangle &= E_\alpha\,\mathbf{H}\,|m\rangle + \boldsymbol{\alpha}\,E_\alpha\,|m\rangle \\ &= (\mathbf{m} + \boldsymbol{\alpha})\,E_\alpha\,|m\rangle \end{aligned}$$

That is, $E_\alpha\,|m\rangle$ is a state with weight $\mathbf{m} + \boldsymbol{\alpha}$. We may denote it by $|m+\alpha\rangle$. We can apply E_α to $E_\alpha\,|m\rangle$ and get a state of weight $\mathbf{m} + 2\boldsymbol{\alpha}$ and so on. Thus, we can construct a series of weights

$$\mathbf{m} + \boldsymbol{\alpha}, \mathbf{m} + 2\,\boldsymbol{\alpha}, \mathbf{m} + 3\,\boldsymbol{\alpha}, \ldots$$

But as the number of weights is finite, the series must terminate after a certain stage. Moreover, we may write

$$E_\alpha\,|m\rangle = f_0\,|m + \alpha\rangle$$

where f_0 is normalization constant. By induction, we have

$$E_\alpha |m + n\alpha> = f_n |m + (n + 1)\alpha>$$

Let us suppose that $\mathbf{m} + k\boldsymbol{\alpha}$ is a weight but $\mathbf{m} + (k + 1)\boldsymbol{\alpha}$ is not a weight. Then

$$E_\alpha |m + k\alpha> = 0$$

or

$$f_k = 0$$

Similarly by the application of the generator $E_{-\alpha}$ on the state $|m>$, we can construct a series of weights

$$\mathbf{m} - \boldsymbol{\alpha}, \mathbf{m} - 2\boldsymbol{\alpha}, \mathbf{m} - 3\boldsymbol{\alpha}, \ldots$$

This series should also terminate after a finite number of terms. As for E_α, we may write here

$$E_{-\alpha} |m - j\alpha> = g_{-j} |m - (j + 1)\alpha>$$

Let us suppose that $\mathbf{m} - j\,\boldsymbol{\alpha}$ is a weight, but $\mathbf{m} - (j + 1)\,\boldsymbol{\alpha}$ is not a weight. Then

$$E_{-\alpha} |m - j\alpha> = 0$$

or

$$g_{-j} = 0$$

Now

$$[E_\alpha, E_{-\alpha}] = r_i(\alpha) H_i = \boldsymbol{\alpha} \cdot \mathbf{H}$$

Operating on the state vector $|m + n\alpha>$, we get

$$\begin{aligned} \boldsymbol{\alpha} \cdot \mathbf{H} |m + n\alpha> &= [E_\alpha, E_{-\alpha}] |m + n\alpha> \\ &= E_\alpha E_{-\alpha} |m + n\alpha> - E_{-\alpha} E_\alpha |m + n\alpha> \\ &= E_\alpha g_n |m + (n - 1)\alpha> - E_{-\alpha} f_n |m + (n + 1)\alpha> \\ &= g_n f_{n-1} |m + n\alpha> - f_n g_{n+1} |m + n\alpha> \end{aligned}$$

or

$$\boldsymbol{\alpha} \cdot (\mathbf{m} + n\,\boldsymbol{\alpha})\,|m + n\,\alpha\rangle = g_n\, f_{n-1}\,|m + n\,\alpha\rangle - f_n\, g_{n+1}\,|m + n\,\alpha\rangle$$

This yields

$$\boldsymbol{\alpha} \cdot \mathbf{m} + n\,\boldsymbol{\alpha} \cdot \boldsymbol{\alpha} = f_{n-1}\, g_n - f_n\, g_{n+1}$$

This is a recursion formula. Varying n from k to – j, we get

$$\boldsymbol{\alpha} \cdot \mathbf{m} + k\,\boldsymbol{\alpha} \cdot \boldsymbol{\alpha} = f_{k-1}\, g_k - f_k\, g_{k+1}$$

$$\boldsymbol{\alpha} \cdot \mathbf{m} + (k - 1)\boldsymbol{\alpha} \cdot \boldsymbol{\alpha} = f_{k-2}\, g_{k-1} - f_{k-1}\, g_k$$

....

$$\boldsymbol{\alpha} \cdot \mathbf{m} + (-j + 1)\,\boldsymbol{\alpha} \cdot \boldsymbol{\alpha} = f_{-j}\, g_{-j+1} - f_{-j+1}\, g_{-j+2}$$

$$\boldsymbol{\alpha} \cdot \mathbf{m} + (-j)\,\boldsymbol{\alpha} \cdot \boldsymbol{\alpha} = f_{-j-1}\, g_{-j} - f_{-j}\, g_{-j+1}$$

These equations are k + j + 1 in number. Adding these together and remembering that $f_k = 0$ and $g_{-j} = 0$, we get

$$(k+j+1)\,\boldsymbol{\alpha} \cdot \mathbf{m} + \frac{1}{2}(k+j+1)\,(k-(k+j)\ \boldsymbol{\alpha} \cdot \boldsymbol{\alpha} = 0$$

or

$$\boldsymbol{\alpha} \cdot \mathbf{m} + \frac{1}{2}(k-j)\ \boldsymbol{\alpha} \cdot \boldsymbol{\alpha} = 0$$

or

$$2\frac{\boldsymbol{\alpha} \cdot \mathbf{m}}{\boldsymbol{\alpha} \cdot \boldsymbol{\alpha}} = j - k = \text{an integer}$$

This proves the first part of the theorem.

But $g(m/\alpha) = 2\alpha \cdot \mathbf{m}/(\alpha \cdot \alpha)$. Therefore, $g(m/\alpha) = j - k$, that is, $-k \le g(m/\alpha) \le j$. Now

$$\mathbf{m} - j\boldsymbol{\alpha},\ \mathbf{m} - (j - 1)\,\boldsymbol{\alpha}, \ldots, \mathbf{m} + (k - 1)\boldsymbol{\alpha},\ \mathbf{m} + k\boldsymbol{\alpha}$$

are the weights, that is, **m** is a weight if $-k \le g \ge j$. Since $g(m/\alpha)$ is also an integer of this range, it follows that $\mathbf{m} - g(m/\alpha)\,\boldsymbol{\alpha}$ is a weight. Obviously, the multiplicity of this weight is the same as that of **m**. ■

Index

A

Abelian group, 4
Abelian subgroup, 15
Abstract group, 27, 31
Adjoint representation, 49, 103
Anti-Hermitian matrix, 92
Antisymmetric matrix, 88
Automorphism, 27

B

Baryon, 172, 176, 177
Basis, 33, 39, 104
Basis vectors, 33–34
  basis of representation, 33
  coordinate system, 33
  scalar, 34
Bilateral symmetry, 152
Binary operation of multiplication, 104
Block matrix, 48

C

Cartan integer, 109
Cartan's theorem, 137
Casimir operators, 165–166
  invariant operator, 165
  problem, 166
  proof, 166
  quantum number, 165
  theorem, 166
Charge conjugation, 159
Class, 17
Clebsch–Gordan series, 59
Closed group, 61
Commutation relations, 105–108
  commutator, 105
  Greek indices, 105
  Hermitian generators, 105
  nonvanishing root, 107
  positive root, 108
  problems, 106
  root vector, 106
  r-parameter semisimple Lie group, 105
  simple positive root, 108
  structure constant, 107
  zero root vectors, 106
Commutator, 81, 101, 105, 197
Compact group, 48, 61
Complex conjugate representation, 49
Complex space, 66, 170
Conjugate classes, 17
Conjugate representation, 49
Conjugate subgroup, 17–18
  elements, 18
  subset, 17
Continuous groups, 61–146
  classification of simple groups, 112–113
    integers, 113
    roots, 112
    tables, 113
    zero angle, 113
  commutation relations between generators of semisimple Lie group, 105–108
    commutator, 105
    Greek indices, 105
    Hermitian generators, 105
    nonvanishing root, 107
    positive root, 108
    problems, 106
    root vector, 106
    r-parameter semisimple Lie group, 105
    simple positive root, 108
    structure constant, 107
    zero root vectors, 106
  computation of highest weight of any irreducible representation of SU(3), 127–131
    example, 129

irreducible representation, 127
nonnegative integer, 128, 130
problem, 131
roots, 127
Weyl's theorem, 128
computation of weights of any irreducible representation of SU(3), 133–135
highest weight of representation, 133
irreducible representation, 133
problem, 134
remark, 134
simple roots, 135
technique for obtaining weights, 134
decomposition of product of two irreducible representations, 139–146
complete column, 144
direct product, 139
double product, 141
elementary particles, 146
first method, 139–141
irreducible representation, 146
Kronecker product, 139, 140, 145
lattice permutation, 143
nonnegative integers, 142
outer product, 139
prescription, 142
second method, 141–146
unimodular unitary group, 141
unitary group, 144
upper limit for first summation, 140
Young's tableaux, 141
YT diagram, 141
definition of continuous group, 61–62
bounded set of numbers, 61
closed set of numbers, 61
compact group, 62
continuous group, 61
discrete group, 61
essential parameters, 61
finite continuous group, 61
mixed continuous group, 62
problem, 62
r-parameter continuous group, 61
dimension of any irreducible representation of SU(N), 131–132
formula, 132
nonnegative integers, 131
generators of Lie groups, 75–84
commutator, 81
example, 78, 80, 82
generator of the group, 77
groups of transformations, 75
identity element, 77
identity transformation, 75, 82
infinitesimal transformation, 75, 83
law of composition, 75
one-dimensional vector space, 81
one-parameter Lie group, 75
problems, 78, 84
r-parameter group, 80
summation convention, 79
transformation equations, 78
transformation matrix, 83
generators and parameterization of group, 98–99
examples, 99
Hermitian generators of Lie group, 99
identity, 99
inhomogeneous linear transformations, 98
generators of SU(2), 91–95
anti-Hermitian matrix, 92
identity transformations, 92
problem, 95
skew-Hermitian matrix, 92
three-parameter group, 94
transformations, 91
unimodular condition, 93
generators of SU(3), 95–98
conditions, 96
diagonal elements, 96
first order of smallness, 96
infinitesimal parameters, 96
problems, 98
transformation matrix, 95
groups of linear transformations, 62–69
Cartesian coordinate system, 63
equivalent matrix equation, 64

general linear group in complex space of n dimensions, 66
general linear group in real space of two dimensions, 66
Hermitian conjugate, 67
homogeneous linear transformation, 63
identity transformation, 66
inhomogeneous linear transformations, 63
law of composition, 67
nonsingular real square matrix, 65
nonsingular transformation, 62
problems, 67, 69
rotation of coordinate system, 64
rotation of position vector in fixed coordinate system, 65
scalars, 62
special linear group, 68
transformation groups, 62
transformation matrix, 63
two-dimensional real space, 63
unimodular unitary group, 68
unitary group, 68
unitary transformation matrices, 67
Lie algebras, 104–105
binary operation of multiplication, 104
commutator, 104
identity, 105
semisimple, 105
significance, 105
vector space, 104
Lie groups, 72–75
analytic function, 75
associative multiplication, 74
identity element, 72
inverse of transformation, 73
Jacobian, 74
law of composition, 72
non-Abelian Lie group, 73
parameter value, 72
problems, 73
matrix representatives of generators, 99–100
Lie group, 99
problems, 100
real group, 100
numerical values of structure constants of SU(3), 122
order of group of transformations, 69–71
diagonal elements, 70
element parameters, 69–70
independent real parameters, 71
Kronecker delta, 70
nondiagonal elements, 71
problem, 71
properties of roots, 108–111
Cartan integer, 109
contradiction, 111
Hermitian operator, 109
nonzero roots, 110
proof, 108, 109
semisimple Lie group, 108
rank of Lie group, 103–104
example, 104
generators, 104
problem, 104
real orthogonal group in two dimensions, 84–91
Abelian Lie group, 91
antisymmetric matrix, 88
connectivity, 90
mixed continuous group, 91
one-parameter group, 89
orthogonal group, 86
orthogonal matrix, definition of, 84
problems, 89, 91
proper rotations, 90
rotation matrix, 87
special orthogonal group, 90
square matrix, 85
transformation equations, 86
two-dimensional real space, 87
roots of SU(2), 114–115
one-dimensional roots, 114
rank, 114
root diagram, 114, 115
roots of SU(3), 115–121
adjacent roots, 115
constants, 116
magnitudes of roots, 117
nonzero roots, 115, 119
orthonormalized roots, 118
problems, 119, 121

root diagram, 119
weight diagram, 120, 121
zero roots, 115
structure constants, 101–103
adjoint representation, 103
commutator, 101
identity, 102
Jacobi identity, 101
regular representation, 103
remark, 103
sufficient conditions, 102
summation convention, 101
structure constants $N_{\alpha\beta}$, 111–112
structure constants, 112
symmetry properties, 111
theorem, 112
weight diagrams, 138–139
remark, 139
root diagram, 139
weights of irreducible representation $D^8(1,1)$ of SU(3), 135–138
Cartan's theorem, 137
fundamental dominant weights, 137
fundamental representations, 138
irreducible representations, 137
maximum multiplicity, 137
nonnegative integers, 138
positive root, 135
series, 136
zero weights, 136
weights of representation, 122–127
dominant weight, 125
eigenvalues, 122
equivalent weights, 125
generators, 122
Hermitian generators, 122, 123
highest weight, 126
irreducible representations, 126, 127
multiplicity of weight, 124
necessary condition, 127
proof, 123, 124, 125
raising operator, 124
remark, 126
simultaneous eigenkets, 125
simultaneous eigenvector, 123
theorem, 123, 124, 125, 126
vector space, 122
weight of eigenstate, 123
Coset, 16–17
Abelian subgroup, 16
collection of elements, 16
definition, 16
left and right, 16
problem, 17
subset, 16
Cyclic group, 14–16
Abelian cyclic subgroup, 15
infinite order, 14
law of composition, 14
matrix group, 15
order of, 14
problems, 14, 15, 16

D

Decomposition, 59, 139
Decuplet, 177
Determinant, 2, 41, 157
Direct product, 27–29, 58, 139
abstract group, 27
arbitrary elements, 29
associative multiplication, 27
automorphism, 27
example, 30
finite groups, 28
identity element, 28
law of composition, 27, 28, 29
problems, 29, 30
subgroup, 28
Discrete group, 61
Dominant weight, 125, 137

E

Eigenstate, 123
Eigenvalue, 60, 165, 171
Eigenvector, 60, 108
Elementary particle, 31, 61
Elementary particles, group theory and, 170–189
baryon decuplet, 177
baryon octet, 176
baryon triplet, 172

fundamental particles, 175
fundamental representation, 170
gauge function, 187
gauge particles, 189
gauge transformation, 184
Gell-Mann–Nishijima formula, 171
Gell-Mann–Okubo mass formula, 177
global gauge symmetry, 186
global gauge transformation, 179
Goldstone boson, 185, 189
ground state, 181
hadrons, 170
Hamiltonian density, 180
Hermitian generators, 170
hidden symmetry, 179
Higgs boson, 189
Higgs mechanism, 185
hypercharge, 170
isospin, 170
Klein–Gordon equation, 185
Kronecker product, 175
Lagrangian density, 178, 179, 181, 188
local gauge symmetry, 185
meson octet, 178
minimum field energy, 182
potential energy, 180, 182
problems, 177
quark–antiquark pair, 174
quark model, 174, 176
Sakata model, 171, 174
unitary gauge, 187
vacuum expectation value, 183
vacuum state, 182

Equivalent matrices, 42

F

Factor group, 19–20
coset multiplication, 19
problem, 19
product of left cosets, 20
quotient group, 20

Field theory, 159
Finite continuous group, 61
First order of smallness, 96
Fully reducible representation, 54

G

Gauge transformation, 178, 185
Gell-Mann–Nishijima formula, 171
Gell-Mann–Okubo mass formula, 177
General linear group, 66
Generator, 14
GL(n), 66, 67, 70
Goldstone boson, 185, 189
Group axioms, 10
Group element, 4–6
example, 5
identity element, 4
problems, 5
unique inverse, 4

Group representations, 31–60
analysis of representations, 51–52
irreducible representations, 52
process, 52
reducible matrix representation, 51
square matrix, 51
basis vectors, 33–34
basis of representation, 33
coordinate system, 33
scalar, 34
change of basis and matrix representative of linear operator, 40–44
equivalent matrices, 42
nonsingular matrix, 42
problem, 44
scalars, 41
similarity transformation, 42, 44
square matrix, 43
transposition of matrix, 41
vector space, 40
complex conjugate and adjoint representations, 49
adjoint representation, 49
complex conjugate representation, 49
construction of representations by addition, 49–50
addition of representations, 50
matrix multiplication, 50
null matrices, 50
square matrices, 49

equivalent and unitary representations, 47
character of matrix, 47
equivalent representations, 47
finite groups, 47
unitary representation, 47
group representations, 44–46
associative multiplication, 46
dimension of representation, 45
faithful representation, 45
identity element, 46
matrix representation, 45
operators, 44
representation space, 44
irreducible invariant subspace, 52
invariant subspace, 52
multiplet of vector space, 52
linearly independent vectors, 33
dimensionality, 33
linear combination, 33
maximum number of, 33
two-dimensional space, 33
linear vector spaces, 31–32
Abelian group, 31, 32
complex vector space, 32
definition, 32
distributive multiplication, 31
field, 31
real vector space, 32
scalars, 31
subspace, 32
matrix representations and invariant subspaces, 52–57
column vector, 56
complementary subspaces, 54
generalized dimensions, 54
group transformations, 55
invariant subspaces, 55, 56
irreducible representations, 55
matrix form, 52
one-dimensional subspace, 53
problem, 57
space decomposed into direct sum, 54
two-dimensional subspace, 54
matrix representative of linear operator, 36–40
example, 39
linear operator, 36, 38
matrix representative, 38
summation convention, 37
operators, 34–35
antilinear operator, 35
equal operators, 35
identity operator, 35
invertible operator, 35
linear operator, 34
nonsingular operator, 35
one-to-one mapping, 34
regular operator, 35
product representations, 57–60
addition of matrix representations, 57
Clebsch–Gordan series, 59
decomposition, 59
direct product, 58, 59
distinct pair, 58
elements of matrices, 57
Kronecker product, 59
Schur's lemma, 60
theorem, 60
reducible and irreducible representations, 48–49
block matrix, 48
decomposable representation, 48
fully reducible representation, 48
square matrix, 48
theorem, 48–49
unitary and Hilbert vector spaces, 35–36
complete unitary vector space, 36
Euclidean space, 36
Hilbert space, 36
inner product, 35
modulus of vector, 35
n-dimensional space, 36
norm of vector, 35
scalar product space, 35
unitary vector space, 35
Group theory, elements of, 1–30
center of group, 18–19
center of G, 18
central normal subgroup, 19
normal subgroup of G, 19
subgroup, 19
characteristics of group elements, 4–6
example, 5

identity element, 4
problems, 5
unique inverse, 4
conjugate elements and conjugate classes, 17
equivalent elements, 17
identity element, 17
problem, 17
conjugate subgroups, 17–18
elements, 18
subset, 17
cosets, 16–17
Abelian subgroup, 16
collection of elements, 16
definition, 16
left and right, 16
problem, 17
subset, 16
cyclic groups, 14–16
Abelian cyclic subgroup, 15
infinite order, 14
law of composition, 14
matrix group, 15
order of, 14
problems, 14, 15, 16
definition of group, 1–4
Abelian group, 4
associative multiplication, 1
binary operation, 1
closure property, 1
element properties, 1
groups, 2
identity element, 2
inverse, 2
law of composition, 1
matrix multiplication, 2
multiplication, 3
order of the group, 4
problem, 3
product, 3
direct product of groups, 27–29
abstract group, 27
associative multiplication, 27
automorphism, 27
finite groups, 28
identity element, 28
law of composition, 27, 28
problems, 29
subgroup, 28
direct product of subgroups, 29–30
arbitrary elements, 29
example, 30
law of composition, 29
problems, 30
elements, 1
factor group, 19–20
coset multiplication, 19
problem, 19
product of left cosets, 20
quotient group, 20
homomorphism, 22–24
identity element, 23
many-to-one mapping, 22
product, 24
properties, 23
theorem, 23–24
identity element, 2
isomorphism, 25–27
automorphism, 27
correspondence, 26
cube roots of unity, 25
identity elements, 26
law of composition, 25
one-to-one mapping, 25
kernel, 24
identity element, 24
mapping, 24
problem, 24
mapping, 20–22
correspondence, 20, 21
examples, 21, 22
image, 21
many-to-one mapping, 21
mapping, 20, 21
one-to-one mapping, 21
prescription 20
unique element, 21
members, 1
multiplication table, 10
associative law, 10
group axioms, 10
normal subgroups, 18
invariant subgroup, 18
nontrivial subgroups, 18
proper subgroups, 18
self-conjugate subgroup, 18
trivial normal subgroups, 18
permutation groups, 6–9

associative multiplication, 7, 8
identity element, 8
inverses, 9
product, 7
symmetric group, 9
writing of, 6
power of group element, 13
analysis, 13
integral powers, 13
positive integer, 13
subgroups, 10–13
binary operation, 11
improper subgroups, 11
index, 13
Lagrange's theorem, 12
law of composition, 10
theorem, 11
trivial subgroups, 11

H

Hamiltonian, 166
Hermitian conjugate, 67, 125
Hermitian operator, 109, 122
Hidden symmetry, 179
Higgs boson, 189
Higgs mechanism, 185
Higgs particle, 189
Hilbert space, 36
Homomorphism, 22–24
identity element, 23
many-to-one mapping, 22
product, 24
properties, 23
theorem, 23–24
Hypercharge, 170

I

Identity element, 2
characteristics, 4
direct product, 28
homomorphism, 23
Lie groups, 72
permutation groups, 8
Index, 13
Infinite group, 4, 61, 91, 161
Infinitesimal transformation, 75, 83, 96
Inner product, 35
Invariant operator, 165
Invariant subgroup, 18
Irreducible representation, 48–49
Isomorphism, 25–27
automorphism, 27
correspondence, 26
cube roots of unity, 25
identity elements, 26
law of composition, 25
one-to-one mapping, 25
Isospin, 170

J

Jacobi identity, 101, 192, 199

K

Kernel, 24
Klein–Gordon equation, 185
Kronecker delta, 70
Kronecker product, 59, 139, 140, 145, 175

L

Lagrangian, 160
Lagrangian density, 160
Lattice permutation, 143
Lie algebra, 104–105
binary operation of multiplication, 104
commutator, 104
identity, 105
semisimple, 105
significance, 105
vector space, 104
Lie group, 72–75
analytic function, 75
associative multiplication, 74
identity element, 72
inverse of transformation, 73
Jacobian, 74
law of composition, 72
non-Abelian Lie group, 73
parameter value, 72
problems, 73
Linear operator, 35
Linear vector space, 31–32
Abelian group, 31, 32
complex vector space, 32

definition, 32
distributive multiplication, 31
field, 31
real vector space, 32
scalars, 31
subspace, 32

M

Mapping, 20–22
correspondence, 20, 21
examples, 21, 22
image, 21
many-to-one mapping, 21
one-to-one mapping, 21
prescription 20
unique element, 21
Matrix, 2, 3
Meson octet, 178
Multiplet, 52
Multiplicity, 124

N

Neutron, 170
Non-Abelian group, 15, 16
Nondiagonal elements, 71
Nonsingular matrix, 2, 42
Nonsingular operator, 35
Nontrivial subgroups, 18
Nonvanishing root, 107
Normalization, 194

O

Observable, 164, 170
Octet, 176, 178
Operator, 34–35
antilinear operator, 35
equal operators, 35
identity operator, 35
invertible operator, 35
linear operator, 34
nonsingular operator, 35
one-to-one mapping, 34
regular operator, 35
Order of a group, 69
Orthogonal group, 86
Outer product, 139

P

Parity, 153–155
Particle, 31, 61, *see also* Elementary particles, group theory and
Permutation groups, 6–9
associative multiplication, 7, 8
identity element, 8
inverses, 9
product, 7
symmetric group, 9
writing of, 6
Physics, *see* Symmetry, Lie groups, and physics
Product representations, 57–60
addition of matrix representations, 57
Clebsch–Gordan series, 59
decomposition, 59
direct product of matrices, 57
distinct pair, 58
elements of matrices, 57
Kronecker product, 59
Schur's lemma, 60
theorem, 60

Q

Quantum field theory, 159
Quark, 174, 176
Quotient group, 20

R

Rank, 103
Reducible representation, 52
Reflection symmetry, 151, 153
Representation, *see* Group representations
Right-handed coordinate system, 154
Root, 106
Root diagram, 115
Root space, 106
Rotation group, 91, 165
Rotation matrix, 87
r-parameter continuous group, 61

S

Sakata model, 171, 174
Schur's lemma, 60

Semisimple group, 18, 165
Semisimple Lie group, 191–194
  commutation relations, 191
  diagonal matrix, 191
  Jacobi's identity, 192
  nonvanishing roots, 193
  normalization, 194
Similarity transformation, 42
Simple group, 18, 112
Skew-Hermitian matrix, 92
Special unitary group, 68
Spontaneous symmetry breaking (SSB), 161
SSB, *see* Spontaneous symmetry breaking
Strong interaction, 164, 166, 178
Structure constant, 101–103
  adjoint representation, 103
  commutator, 101
  computation of, 199–201
  identity, 102
  Jacobi identity, 101, 199
  regular representation, 103
  remark, 103
  sufficient conditions, 102
  summation convention, 101
  symmetry relations, 200
SU(3), 95–98
  conditions, 96
  constants, 116
  diagonal elements, 96
  first order of smallness, 96
  infinitesimal parameters, 96
  magnitudes of roots, 117
  orthonormalized roots, 118
  problems, 98, 119, 121
  root diagram, 119
  transformation matrix, 95
  weight diagram, 120, 121
Subgroup, 10–13, 18
  binary operation, 11
  improper subgroups, 11
  index, 13
  invariant subgroup, 18
  Lagrange's theorem, 12
  law of composition, 10
  nontrivial subgroups, 18
  proper subgroups, 18
  self-conjugate subgroup, 18
  theorem, 11
  trivial normal subgroups, 18
  trivial subgroups, 11
Symmetric group, 9
Symmetry, Lie groups, and physics, 147–190
  Casimir operators, 165–166
    invariant operator, 165
    problem, 166
    proof, 166
    quantum number, 165
    theorem, 166
  group theory and elementary particles, 170–189
    baryon decuplet, 177
    baryon octet, 176
    baryon triplet, 172
    fundamental particles, 175
    fundamental representation, 170
    gauge function, 187
    gauge particles, 189
    gauge transformation, 184
    Gell-Mann–Nishijima formula, 171
    Gell-Mann–Okubo mass formula, 177
    global gauge symmetry, 186
    global gauge transformation, 179
    Goldstone boson, 185, 189
    ground state, 181
    hadrons, 170
    Hamiltonian density, 180
    Hermitian generators, 170
    hidden symmetry, 179
    Higgs boson, 189
    Higgs mechanism, 185
    hypercharge, 170
    isospin, 170
    Klein–Gordon equation, 185
    Kronecker product, 175
    Lagrangian density, 178, 179, 181, 188
    local gauge symmetry, 185
    meson octet, 178
    minimum field energy, 182
    potential energy, 180, 182
    problems, 177
    quark–antiquark pair, 174
    quark model, 174, 176

Sakata model, 171, 174
unitary gauge, 187
vacuum expectation value, 183
vacuum state, 182
symmetry, 147–165
axis of symmetry, 148, 150
bilateral symmetry, 152
charge conjugation, 159–162
combination of symmetry operations, 155–156
conservative system, 158
continuous symmetry, 150
discrete summary, 148
double mirror, 152
dynamical symmetries, 159
external symmetries, 159
fourfold point of rotational symmetry, 147
Galilean transformations, 162, 163
geometrical symmetries, 159
Hermitian generators, 164
higher and lower symmetries, 151
homogeneity of time, 159
homogeneous space, 157
identity operation, 148
improper axis of rotation, 156
improper rotational symmetry, 156
infinite pattern, 156
internal symmetries, 159
invariant square, 147
inversion, 152
irreducible representation, 164
isotropic space, 150
Lagrangian density, 160
Lorentz transformation, 157, 161, 163
Maxwell's equations, 164
multiple symmetries, 155
Newton's equations of motion, 160
Newton's second law of motion, 157
parity, 153–155
plane of reflection symmetry, 152
point of fourfold rotational symmetry, 147
potential energy, 154
problem, 163
proper rotation, 155
reflection/inversion symmetry, 151–152
remark, 155
right-handed coordinate system, 154
rotational symmetry, 147–151, 161
Schrödinger wave equation, 158
space–time symmetries, 159
spontaneous symmetry breaking, 161
square possessing two types of symmetries, 156
symmetry element, 147
symmetry groups and physics, 162–165
symmetry operation, 147
time-reversal symmetry, 157–159
translational symmetry in space, 156–157
unitary symmetry, 166
symmetry and physics, 166–169
example, 168
Hamiltonian operator, 167
Hermitian generators, 167
proof, 167, 168, 169
semisimple Lie group, 166
theorem, 167, 168, 169

T

Theorem
Cartan's, 137
Casimir operators, 166
homomorphism, 23–24
Lagrange's, 12
product representations, 60
reducible and irreducible representations, 48–49
subgroups, 11
symmetry and physics, 167
weights of representation, 123
Weyl's, 128
Time reversal, 157
Translation group, 90
Trivial subgroups, 11

U

Unitary gauge, 187
Unitary group, 68, 170
Unitary operator, 99, 167
Unitary representation, 47
Unitary symmetry, 166

V

Vacuum expectation value (VEV), 183
Vector space
  linear, 31–32
  multiplet of, 52
  one-dimensional, 81
  unitary and Hilbert, 35–36
  weights of representation, 122
VEV, *see* Vacuum expectation value

W

Weak interaction, 155, 164
Weight, 122–127
  dominant weight, 125
  eigenvalues, 122
  equivalent weights, 125
  generators, 122
  Hermitian generators, 122, 123
  highest weight, 126
  irreducible representations, 126, 127
  multiplicity of weight, 124
  necessary condition, 127
  proof, 123, 124, 125
  raising operator, 124
  remark, 126
  simultaneous eigenkets, 125
  simultaneous eigenvector, 123
  theorem, 123, 124, 125, 126
  vector space, 122
  weight of eigenstate, 123
Weight diagram, 120, 138–139
  remark, 139
  root diagram, 139
Weight space, 123
Weyl's theorem, 128

Y

Young's tableaux, 141